Environmental Deterioration of Materials

WITPRESS

WIT Press publishes leading books in Science and Technology.
Visit our website for the current list of titles.
www.witpress.com

WITeLibrary

Home of the Transactions of the Wessex Institute, the WIT electronic-library provides the
international scientific community with immediate and permanent access to individual
papers presented at WIT conferences. Visit the WIT eLibrary at
http://library.witpress.com

International Series on Advances in Architecture

Objectives

The field of architecture has experienced considerable advances in the last few years, many of them connected with new methods and processes, the development of faster and better computer systems and a new interest in our architectural heritage. It is to bring such advances to the attention of the international community that this book series has been established. The object of the series is to publish state-of-the-art information on architectural topics with particular reference to advances in new fields, such as virtual architecture, intelligent systems, novel structural forms, material technology and applications, restoration techniques, movable and lightweight structures, high rise buildings, architectural acoustics, leisure structures, intelligent buildings and other original developments. The Advances in Architecture series consists of a few volumes per year, each under the editorship - by invitation only - of an outstanding architect or researcher. This commitment is backed by an illustrious Editorial Board. Volumes in the Series cover areas of current interest or active research and include contributions by leaders in the field.

Associate Editors

Environmental Deterioration of Materials

Editor

A. Moncmanová
Slovak Technical University, Slovakia

WITPRESS Southampton, Boston

Editor

A. Moncmanová
Slovak Technical University, Slovakia

Published by

WIT Press

Ashurst Lodge, Ashurst, Southampton, SO40 7AA, UK
Tel: 44 (0) 238 029 3223; Fax: 44 (0) 238 029 2853
E-Mail: witpress@witpress.com
http://www.witpress.com

For USA, Canada and Mexico

WIT Press

25 Bridge Street, Billerica, MA 01821, USA
Tel: 978 667 5841; Fax: 978 667 7582
E-Mail: infousa@witpress.com
http://www.witpress.com

British Library Cataloguing-in-Publication Data

A Catalogue record for this book is available
from the British Library

ISBN-10: 1-84564-032-2
ISBN-13: 978-1-84564-032-3
ISSN: 1368-1435

LOC: 2006929309

The text of the chapters of this book were set individually by the authors and under their supervision.

No responsibility is assumed by the Publisher, the Editors and Authors for any injury and/or damage to persons or property as a matter of products liability, negligence or otherwise, or from any use or operation of any methods, products, instructions or ideas contained in the material herein.

Contents

Chapter 3
Corrosivity of atmospheres – derivation and use of information
D. Knotkova & K. Krieslova

Chapter 4
Atmospheric corrosion and conservation of copper and bronze
D. Knotkova & K. Krieslova

Chapter 5
Environmental deterioration of concrete
A.J. Boyd & J. Skalny

Chapter 6
Corrosion of steel reinforcement
C. Andrade

Chapter 7
Environmental pollution effects on other building materials
P. Rovaníková

Chapter 8
Environmental deterioration of building materials
R. Drochytka & V. Petránek

Chapter 9
Environmental deterioration of timber
S. Mindess

Preface

The increasing level of pollution in the environment not only harms the natural world, but also accelerates the deterioration and corrosion of materials used in technical work as well as materials of objects with historical or artistic value. It is not possible to eliminate the numerous sources of this negative effect, so there is a currently an increased effort at better preservation, which requires a thorough knowledge of the causes of the degradation of individual materials.

The purpose of this book is to evaluate various natural and anthropogenic factors, to better understand the conditions that cause the decay of selected construction materials, and to show conservation approaches and application methods allowing for their improved preservation. The chapters of the book offer an overview of the newest information regarding the causes of material and object deterioration and its prevention for a wide specialized readership. The authors of these chapters have based their work on knowledge gained from years of research and practical experience during the assessment of damaged objects. The content of the book is relatively wide but some topics are not involved. Topics such as flood damage, seismic effects, mechanical vibrations due to transportation, and indoor building environments, among others, are not covered, or are given limited attention. Only those construction materials that are used most often or to the greatest extent have been selected. Nevertheless, the book offers the specialized reader a general introduction to many facets of material preservation, and may be a good supplement to the limited publications pertaining to this field.

I would like to express my thanks to all of the contributors for their patience, their cooperation and their support over an extended period of time. I would also like to thank the publishers for their support and cooperation. I believe that the final result of our efforts is a worthwhile contribution to the better understanding of the environmental deterioration of materials.

A. Moncmanová
Slovak Republic
2006

Introduction

A. Moncmanová
*Faculty of Chemical and Food Technology, Slovak University of
Technology, Bratislava, Slovak Republic.*

This book deals with the basic principles underlying the environmental deteriora-tion of commonly used construction materials and metals, as well as the available prevention and protection techniques that may be adopted to protect and conserve them. Building and construction materials, such as natural stones and timber, man-made composite materials, such as bricks and concrete blocks, and construction metals, namely, cast iron, steel, aluminium, zinc and copper and their alloys, are included in this book. The individual chapters include brief information about some types of polymer composites and paints. Taking into consideration the wide range of these materials used in construction, more detailed information on the factors influencing the failures of individual types of polymeric materials is beyond the scope of this book.

This book briefly reviews the basic knowledge on the environmental deteriora-tion mechanisms of materials and corrosion of metals and the information on the meteorological and climatic factors, the effects of pollutants and biological action affecting degradation. This also includes information about ISO standards relating to the classification of atmospheric corrosivity, the classification system of low corrosivity of indoor atmospheres and corrosivity categories of atmospheres for several construction metals. Dose–response functions describing the corrosivity of outdoor atmospheres for these metals are also briefly discussed. The book gives several examples of how this knowledge may be applied to the protection and con-servation of buildings, industrial facilities and culturally important objects in order to reduce damages and economic impact.

The deterioration of buildings and the corrosion of metallic constructions due to natural environmental effects and anthropogenic pollution lead to shortening of the service life and usefulness of structures, demand the use of more valuable and expensive materials, give rise to higher costs of surface protection or replacing damaged segments, thus having serious economic impact. The depreciation of construction materials has a severe effect on the national economies. A number of reports focusing on the corrosion costs in several countries were written in the last century. They indicated that the total annual corrosion cost varies between 1.5% and 5.2% of the gross national product of the respective economies. In the last American study, the total direct cost of corrosion was estimated to be 3.1% of the 1998 US gross domestic product (GDP), the indirect cost of corrosion was estimated to be equal to the direct cost and the sum of both represents approximately 6% of GDP. In the industrially polluted regions, indirect losses may be considerably higher than the direct cost and give rise to significantly higher economic losses. Air pollution also has a considerable impact on the deterioration of stone and concrete buildings in polluted industrial areas. The degradation of slab blocks in the large housing estates built in the extremely polluted cities of Central and Eastern Europe has been noted over the past 40 years. The economic impact of the deterioration of construction timber is also serious and requires a considerable investment to maintain or replace parts of damaged timber. Annually, the global losses incurred by the decay of wood through biological processes have been estimated at US$ 10 billion.

Irreplaceable losses may result when air pollution affects historical sites, stone statuary and other works of art. This deterioration is related not only to physical damage but also to the aesthetic appeal of the material surface. Air pollution damage to artefacts has been known since the Middle Ages. Concerns about these effects on materials were raised as early as 1284, when a Royal Commission was appointed to study air pollution from fuel used for kilns in London. By the 17th century, many references were made to the degradation of materials and products in polluted atmosphere; however, it was only in the 20th century that attention began to be paid to the better understanding of the causes of deterioration and to the effective prevention and protection of culturally important artefacts and monuments against deterioration.

1 Chemical, physical and biological factors affecting the deterioration of materials

The degradation of materials is the result of different physical, chemical and biological effects. Deterioration also depends on the type of materials and products used and is influenced by specific conditions of processing and use. Physical degradation takes place due to mechanical attrition of the material surface, embrittlement, failure of the component by breaking due to fatigue stress or other irreversible changes of the shape; a specific example of physical degradation is the interaction of radioactive radiation with material in a nuclear power station. For example, sorption of the well-known pollutant sulphur dioxide on a stone surface induces physical changes such as water retention and change in porosity. The repeated freezing and thawing process may result in non-porous layer generation and consequently in blistering and splitting-off of the layer. Materials deteriorate due to chemical attack through exposure to different aggressive chemical compounds. The presence of anthropogenic pollutants in the environment can accelerate this deterioration. The interaction of physical and chemical factors can significantly increase the degradation of materials. The effect of condensation of moisture on the surface of materials and the absorption of acidic gases in generating aqueous film with subsequent formation of aggressive compounds during chemical reactions and the effect of the volume change due to chemical reactions with contaminants in concrete pores are examples.

It is assumed that damage to stone and concrete is predominantly associated with natural environmental factors and, to a lesser extent, with pollution. Timber degrades due to chemical and physical exposure and the effect of various types of biological agents including insects, bacteria and fungi. Corrosion of metals is affected by the nature of the substances present in the environment including moisture and oxygen and also by biological agents. In a dry atmosphere, the corrosion of metals is negligible at normal temperature; it progresses only in the presence of moisture on the metal surface, where the generated moisture layer provides a medium for water-soluble air pollutants and is a conducting medium for corrosion process reactions. Many metals form a protective corrosion layer that guards against corrosion, but higher concentrations of pollutants decrease this protective effect. The properties of the protective corrosion layer differ with the type of metal. Long-term corrosion resistance of some metals such as steel appears to depend on changes in the structure and composition of the protective corrosion layer.

Deterioration of building materials and corrosion of metals occur not only in the natural environment, e.g. in air, water and soil, but also in environments polluted by industrial, agricultural and municipal wastes in the form of gases, solutions and solids. This damage may be very unfavourable with respect to the service life of structures and their reliability in operation.

Materials also undergo degradation in the absence of external factors through the effects of the structure failure of the material and changes in the physical and chemical properties of the material. The characteristics and properties of the material are as equally important as the environmental factors in the process of deterioration. The principal characteristics of most material damage mechanisms are the texture of the material surface, the chemical properties of the surface, the specific surface of the exposed area and the medium in which the material is found. Especially important is the roughness of materials; rough surfaces permit easier mass transfer across the quasi-laminar air film close to the surface. The rate of delivery of gases to the surface is largely determined by the mutual chemical affinity of the material of the surface and the particular gas. Irregularities in the structure significantly influence level of weathering of the material surface. The corrosion of metals also depends on, besides other factors, the properties of surface electrolytes. The most important property of stones is porosity, in terms of the number, size and extent to which pores are connected. It can in turn determine, for example, the amount of salt solution that can be absorbed and the rate of its migration through the stone. In this manner, the material's sensitivity to deterioration will be enhanced. For instance, the presence of several interconnecting microspores that can carry moisture into the stone make it susceptible to frost and pollution attack. Other material properties such as permeability and absorbability may affect its resistance considerably. The permeability of a building material is conditioned by its porosity – the absorbability may decrease the material's resistance to frost and influence the weathering of materials such as stone, concrete, brick and plaster; however, it can also be utilised for protection, e.g. in the empty-cell process used for the treatment of timber. Generally, the properties and structure of a material may not only significantly affect its quality but also influence its extent of damage. The extent of the effects depends on the quality of the raw materials used to produce the product and the mode of processing. It is associated with the way in which the material is incorporated into the final product, the correct preservation and maintenance of the product, and the way it is designed and used.

The knowledge of the impact of natural environmental and anthropogenic factors together with the effects of non-external factors on construction and building materials is important for the better understanding of deterioration mechanisms and for the optimal selection of materials, and is also a basis for the design of materials and for the proper selection of prevention and protection measures.

Principal sources and literature for further reading

[1]   Baer, N.S. & Banks, P.N., Indoor air pollution: effects on cultural and historical materials. *The International Journal of Museum Management and Curatorship*, **4**, pp. 9–20, 1985.
[2]   Koch, G.H., Brongers, M.P.H., Thompson, N.G., Virmani, Y.P. & Payer, J.H., *Corrosion Costs and Preventive Strategies in the United States*, Report No. FHWA-RD-01-156, CC Technical Laboratories, Inc.: Dublin, OH, 2001.
[3]   Lipfert, F.W., Benarie, M. & Daum, M.L., *Derivation of Metallic Corrosion Damage Functions for Use in Environmental Assessment*, BNL 35668, Brookhaven National Laboratory: Upton, NY, 1985.

CHAPTER 1

Environmental factors that influence the deterioration of materials

A. Moncmanová

Faculty of Chemical and Food Technology, Slovak University of Technology, Bratislava, Slovak Republic.

1 Introduction

The process of deterioration of materials induced by outdoor environmental factors is a complex interplay of the effects of climate and local meteorological characteristics, of biological processes and frequently of complicated chemical processes resulting from the impact of pollutants and natural constituents from the surrounding environment. Except for the decay caused by cataclysmic events, the principal natural environmental factors affecting the deterioration of materials include, but are not limited to, moisture, temperature, solar radiation, air movement and pressure, precipitation, chemical and biochemical attack, and intrusion by micro- and macro-organisms. Natural factors, together with those expected to be a part of the pollution processes, continuously promote weathering and material decay, including metal corrosion.

The evaluation of the influence of each of the environmental factors in a given situation requires an understanding of which mechanisms are potentially of concern for the type of material or structure in question. In general, the main deterioration mechanisms include: (1) erosion, (2) volume change of the material and the volume changes of the material in pores, (3) dissolution of a material and the associated chemical changes, and (4) biological processes.

In material science, erosion is described as the recession of a surface because of repeated localized shock. In an outdoor environment, the erosion of a material is usually affected by suspended abrasive particles, especially by fine solid particles driven against the material surface by moving fluids. The surface area to particle size ratio controls deterioration because for a given volume of material, the smaller particle size, the larger is the surface area exposed to erosion (the presence of joints also increases a material's surface area). The nature of the fluid flow is likewise important – a markedly smaller effect may be observed for laminar flow than for turbulent flow.

Changes in the volume of a material are a function of temperature, solar radiation and humidity; the volume changes of the material in pores are controlled by humidity and temperature. The thermal expansion and contraction of a material is influenced by temperature variation, which, in outdoor environment depends on the duration and intensity of exposure to the incoming solar radiation and the direction in which the material surface faces (although expansion caused by changes in temperature from day to night or from summer to winter has only a minor effect). Rainfall, fog and wind can cause volume variation due to uneven moisture content. The differential dimensional changes associated with the change in moisture content can be observed in different materials that are bonded together. Such dimensional changes can have a warping effect. A similar attack may be caused by differential moisture content through the layer of a homogeneous material, since the side with a lower moisture content will expand less than that with a higher moisture content (this effect can be observed when rainwater is absorbed on the outdoor surface of a material). Water in combination with freezing conditions can lead to a rapid destruction of certain materials. Water that penetrates into a material may freeze. Frozen water expands and strong pressure is exerted on the structure of the material. A climate with temperature fluctuations across the freezing point can cause the most damage, as repeated freezing cycles can effectively destroy the surface material. The degradation process thus depends upon the degree of saturation with water, the rate and number of freezing cycles, the elastic properties and the strength of the material, and the pore structure of the material.

Deposited contaminants and products of reaction between pollutants and materials can also induce expansion. Chemicals from the surrounding environment can affect the oxidation of a material into more voluminous forms; surface hygroscopic contaminants such as salts, metal oxides, and vegetal fibres in which moisture can be present also play an important role. They can absorb water in different ways including: binding using surface energy, sorption with the formation of a hydrate, diffusion of water molecules in the material structure, capillary condensation, and formation of solution. The moisture content of these substances depends on the nature of the substance, the temperature, and the partial pressure of water vapour in the immediate environment.

Chemically induced damage involves dissolution, oxidation, and hydrolysis. The decay is a result of the interaction between material and natural constituents and pollutants and the amount of water present. The interactions will vary depending on the reactivity of a material, the character of the intercepting surface, the extent of exposure, and the nature of the contaminants. The chemical changes are enhanced by heat (most chemical reactions proceed more rapidly as the temperature increases); therefore, chemical damage to materials is most prevalent in warm, humid climates.

Dissolution of building materials, especially structures containing carbonates, is most frequently caused by the action of acidic solutions such as rain containing carbonic acid or both carbonic acid and sulphuric acid. The overall effect of carbonic acid and sulphuric acid can be extensive weathering of a surface. The crystallization of dissolved compounds from solutions and the hydration of deposited contaminants and products formed during the reaction of a material with chemicals accompany crystallization and hydration pressures. The pressures may be surprisingly high and can lead to mechanical deterioration of a structure.

Oxidation of a material by atmospheric oxygen leads to chemical changes in the composition of the material (especially the material surface); for example, the reaction of metal ions with oxygen to form oxides or hydroxides.

Hydrolysis reaction between a material and water can affect the dissolution of the material and/or change of its chemical composition. Hydrolysis leads to a decrease in material cohesion and a change in porous structure.

The biological factors that cause deterioration include biochemical effects and intrusion by different organisms. A biochemical attack is usually a crucial factor in the biodeterioration of building structures, as the metabolites (such as enzymes, excrements, and faeces) of macro-organisms, plants, and animals living in the material can affect the material chemically. Hyphae of fungi and lichens and plants root systems, which spread through the structures, induce mechanical damage of the structures. In addition, boring insects may destroy structure cohesion, which can allow water to penetrate more quickly and deeply, speeding other deterioration processes.

2 Meteorological and climatic factors

Climatic conditions and meteorological characteristics can affect damaging processes such as mechanical stress, desiccation, surface scaling, attrition, and cracking and may accelerate certain forms of chemical attack on materials.

Climatic factors including air temperature, sunlight radiation, air humidity, different forms of precipitation (rain, snow, etc.), and wind velocity and direction are undoubtedly significant. The impact of these factors varies and depends on the seasons and the intraseasonal variability because they determine the duration of sunshine, the temperature fluctuation (e.g. rapid and large diurnal temperature changes), air movement/pressure, rain intensity and frequency, and local soil hydrology.

The meteorological conditions affect processes of transport, transformation, dispersion, and deposition of emissions from sources and thus may influence pollution-induced damage of materials. Conversely, the cycles of pollutants can affect physical processes in the atmosphere. Increasing atmospheric concentrations of carbon dioxide and other trace gases alter air chemistry and affect chemical reactions. Carbon dioxide emissions influence temperature variation by the greenhouse effect, which can accelerate reactions and stimulate chemical changes on the material surface in an outdoor environment. Sulphur dioxide emissions have a cooling effect as they backscatter sunlight and produce brighter clouds by allowing smaller water droplets to form.

2.1 Moisture

Moisture and temperature affect the chemical, biological, and mechanical processes of decay. The formation of a moisture layer on the material surface is dependent upon precipitation. It may also be generated as a result of the reaction of adsorbed water with the material surface, deposited particles with the material surface, and deposited particles with reactive gases.

On a surface with moisture, particles have a better possibility of adhering and water-soluble gases are more readily captured. Both gas and particle fluxes increase when condensation takes place on the material surface and decrease when evaporation occurs.

A moisture layer is a medium for the chemical and photochemical reactions of surface contaminants and is also a conductive path for the electrochemical reactions. Two variables are important from the viewpoint of the damage caused by moisture: dew point and relative humidity of air. Dew point is a characteristic of the water content of the large-scale air mass, whereas relative humidity depends on the local temperature and therefore on the local meteorological parameters. When the temperature of a material is below the ambient dew point, water condenses on the material, a moisture layer (condensation) can form, and the material damage may proceed. In most materials, an increase in relative humidity causes further deterioration due to more prolonged wetness time, higher deposition rates of pollutants and better conditions for biodeterioration.

Among climatic factors, humidity plays the most important role in outdoor metal corrosion. In the absence of atmospheric moisture, there will be very limited non-pollutant-induced and pollutant-induced corrosion. The rate and nature of the corrosion is a function of relative humidity, sunlight radiation, surface contaminants, the properties of the film of electrolytes formed on the metal surface, and the duration of the effect on the metallic surface. The rate of drying after wetting is dependent on conditions such as ambient temperature, relative humidity, and wind velocity. Corrosion induced by pollutants results from the interaction of a particular pollutant with the metal surface and the metal corrosion layer.

For metallic corrosion, the critical relative humidity value and the time-of-wetness (TOW) are defined. The critical relative humidity value is the minimum concentration of water vapour required for corrosion to proceed. The film of electrolytes is formed at a certain critical relative humidity level. The critical relative humidity value depends upon the material that is being corroded, the hygroscopic nature of the corrosion products and surface deposits, and the nature and level of atmospheric pollutants present. Hydrated products and hygroscopic salts can decrease the 'critical relative humidity value', resulting in large amounts of moisture on the metal surface.

TOW refers to the period of time during which the atmospheric conditions are favourable for the formation of a layer of moisture on the surface of a metal (or alloy). The TOW value may be indicated experimentally or estimated from the temperature–humidity complex. Instrumentally recorded TOW can be determined by various measuring systems and is strongly correlated to the number of days with precipitation and average temperature above freezing. For the purpose of standards, TOW is defined as the time period during which the relative humidity is in excess of 80% and the temperature is above 0°C. Changes in seasonal and annual rainfall will change the period of time for which a surface is wet, affecting the surface leaching and the moisture balance that influence material decay processes.

Moisture is considered to cause significant damage to inorganic building materials. The moisture content and the permeability predominantly influence the impact of pollutants on the rate of weathering of the material. The main risk factor is the cyclic variation of moisture content (wetting and drying) in the presence of hygroscopic salts. The rate of drying after wetting is dependent on conditions such as ambient temperature, relative humidity, and wind velocity. Deterioration of stone proceeds usually at higher relative humidity, above 65%, and is frequently associated with freeze–thaw weathering. The deterioration of external walls made of porous building materials is caused by the excessive moisture content, particularly after driving rain and exposure to long durations of moist conditions.

2.2 Temperature

Temperature affects the processes of deterioration of a material gradually and in a variety of ways. Changes in temperature induce a thermal gradient between the surface layer and the inner layer of materials (particularly in materials with lower thermal conductivity), which may result in the degradation of the mechanical properties of the material and can lead to the formation of fine cracks. The formation of cracks is accompanied by a loss of strength and by an increase in material porosity, which may lower the chemical resistance of the material.

Temperature fluctuation of materials may influence bulk expansion, such as the expansion of stone grains, the dilatation of different materials in joints, and the expansion of water in material capillaries.

Increased ambient air temperature is one of the reasons why the rate of wet deposition is more important for deterioration processes in tropical and subtropical regions than in temperate regions. Higher ambient temperature reduces the effects of freeze–thaw cycles.

Large thermal stress may originate during a fire, when temperature rapidly increases within a short time and may induce serious damage to specific types of materials (e.g. explosive spalling in concrete – see Section 4.2 in Chapter 8).

In a polluted environment, a rise in ambient air temperature can speed up material damage due to the increased rate of chemical reactions on the material surface. However, lowered ambient temperature may enhance the chance of damage. During a temperature inversion, objects usually cool to a temperature below that of the surrounding air. If the object's surface temperature falls below the dew point, a condensation layer will form on the surface and, in the presence of pollutants (whose concentrations may increase under the influence of the temperature inversion), this will accelerate deterioration.

2.3 Solar radiation

Solar radiation causes temperature changes in materials and may induce volume changes of material in the pores due to expansion of water which is heated by solar radiation. Solar radiation plays an important role in photochemical reactions since it supplies the energy for the excitation or splitting of bonds in the reacting molecules. Adequate intensity of solar radiation at suitable wavelengths is an essential condition for photochemical reactions that influence the deterioration of different construction materials. Of the total energy impinging on the earth's surface, approximately 10% is ultraviolet radiation, 45% is visible light, and 45% is infrared radiation. The primary substances absorbing ultraviolet radiation include molecular oxygen, oxides of sulphur and nitrogen, and organic compounds such as aldehydes. A molecule of oxygen is split into reactive oxygen atoms after the absorption of ultraviolet radiation below 240 nm. Sulphur dioxide reaches the excited state only during the absorption of solar radiation at wavelengths of 340–400 nm. The oxidation of excited sulphur dioxide in air leads to the generation of sulphur trioxide. Sulphur trioxide subsequently reacts with ambient moisture to form sulphuric acid aerosol. In contrast to sulphur dioxide, nitrogen dioxide is photochemically very active. It absorbs solar radiation over the whole of the visible and ultraviolet range. In a spectral range of 380–600 nm, excited molecules are formed, and below 43 nm photodissociation occurs. Sunlight-induced atmospheric photochemical reactions between nitrogen oxides, organic gases, and vapour produce ozone as a reaction product. Ozone accelerates the deterioration of synthetic organic materials used as protective coatings against environmental effects on a variety of surfaces including metals, concrete, wood, etc. Ozone was also found to have an incentive effect on limestone and dolomitic sandstones. Visible and ultraviolet radiation result in the generation of free radicals in wood. The oxidation of these radicals leads to the formation of oxidation products or to the decomposition of radicals, which is accompanied by the production of low molecular weight products.

2.4 Wind effect

An increase in wind velocity may affect the deterioration of materials in various ways. Wind drives liquid and solid particles from the air to the material surface, where they cause local attrition and contribute to the weathering of the material. The kinetic energy of particles and the degree of inertial impaction of droplets on the material surface depend on the wind velocity. The high-speed end of the wind spectrum is of interest for abrasion and the low-speed end for diffusion. Wind flows around buildings can influence the deposition rates of both gaseous and particulate pollutants, as well as strengthen the effect of driving rain. During rainfall, windward walls are wetted considerably more than leeward walls. On the windward side, wind promotes the penetration of rainwater and aqueous solutions into porous materials or, during other weather conditions (solar radiation, temperature changes), supports desiccation of the aboveground part of a construction. Both processes, wetting and drying, may affect volume changes in the structure. A serious effect of wind may be an increasing transport of sea salt inland which can substantially extend the areas that are affected by marine aerosols along seacoasts.

3 Harmful effects of air, water, and soil

3.1 Effects of air pollutants and natural atmospheric constituents

In the atmosphere, materials undergo deterioration through a series of physical and chemical interactions and biological activity. In addition to environmental loading, exposure to atmospheric pollutants accelerates the damage and multiplies the effect of weathering.

The atmosphere is a mixture of gases and liquid and solid particles. Dry air contains 18 permanent gaseous constituents. Among these, nitrogen, molecular oxygen, argon, and carbon dioxide are the major constituents, whereas the rest are gases present in minor quantities. The minor constituents include the group of trace gases that are considered to be pollutants: nitrous oxide, carbon monoxide, ammonia, ozone, nitric oxide, nitrogen dioxide, sulphur dioxide, and hydrogen sulphide. Particulate matter of biological origin, dust, small crystals, and droplets represent solid and liquid atmospheric substances. Water can be found as water vapour as well as in liquid and solid phases.

The atmosphere is polluted both by the release of different chemicals into the air (primary pollutants) and by chemical processes occurring directly in the air (secondary pollutants). Primary pollutants enter the atmosphere from natural sources and due to human activity. In the atmosphere, they react with water vapour and water droplets, oxygen, and other reactive substances such as liquid and solid particles and form secondary pollutants. Among the pollutants that cause damage, sulphur and nitrogen compounds and particles deserve special mention. It has been demonstrated clearly that they enhance the natural weathering of various materials but in some cases it may be difficult to quantify their contribution to this process.

In the atmosphere, primary and secondary pollutants are transported, scattered, and deposited on the surface of a material. On the material's surface, pollutants induce the reactions either with the material or with the reactive products generated on the surface. The extent of exposure, the chemical composition, and the concentration of pollutants, coupled with the amount of moisture present on the surface, determine the changes in the material and the extent of deterioration. The structure and the type of material are likewise important. Along with these, the atmospheric corrosion of metals is also dependent upon other factors such as concentration of surface electrolytes and variability in electrochemical reactions, the orientation of the metal surface, and the influence of the protective corrosion layer.

3.1.1 Wet and dry deposition of pollutants
Different substances that are released into the air or those that originate during the processes occurring in air may fall to the ground in either dry form (*dry deposition*) or wet form (*wet deposition*). The sum of both, the dry and wet depositions,

is *total deposition*. It is estimated that dry deposition ranges from 20% to 60% and wet deposition varies from 80% to 40% of the total deposition.

Wet deposition refers to substances that are scavenged from the air by hydrometeors such as rain, clouds, or fog drops. Dry deposition relates to particles and gases that fall on the material surface close to the sources of emissions. Wet deposition loads sudden but random doses of contaminants (most of which are usually in a dilute solution), whereas dry deposition is a slower but more continuous process. Wet deposition can influence areas that are many tens of kilometres away from the sources of pollution; however, under certain meteorological conditions, the dry deposition process is also capable of delivering pollutants to areas that are considerably distant from the pollution sources. It is assumed that within highly polluted industrial areas, dry deposition delivers as much material to exposed surfaces as does wet deposition and is relatively more important.

Factors influencing *dry deposition* are meteorological conditions, the nature of the material's surface, and the chemical and physical properties of the deposited species. During dry deposition substances are transported downwards by turbulent diffusion, eddy transport, or sedimentation through the atmospheric surface layer. The mechanisms that take place close to the surface in a region of calm are gravitational settling, inertial impaction and interception or Brownian motion for particles as well as diffusion processes for gases. The temperature and humidity gradient near the surface can also support or retard the deposition of both gases and particles. Thermophoresis (in the submicron size range) can interfere with Brownian deposition when the surface is warmer than the air. Electrophoresis is another deposition mechanism that is active in the presence of an electric field.

The deposition of gases on the surface of a material is to a large extent determined by the chemical affinity of the surface, the extent of exposure, and the chemical composition and reactivity of gases in the vicinity of the surface that control the reaction between the gases and the intercepting surface. Atmospheric turbulence plays an important role in particle deposition because it determines the rate at which particles are deposited on the surface. Particle deposition is strongly influenced by particle size. Inertial impaction and interception are usually highly efficient for particles larger than $10\,\mu m$ in size and can be enhanced by gravitational settling. Impaction is not effective for particles of size less than $0.3\,\mu m$ due to their low inertia. Brownian motion transports submicron particles, but Brownian diffusion can be so low that small particles have difficulty penetrating into the surface.

Wet deposition is more complex than dry deposition due to the great variety of processes and parameters influencing this action (equilibrium chemistry, more phases and phase transitions, different chemical reactions or transformations in the water phase, etc.). In addition to these, the process can take place either inside or outside a cloud, and pollutants might exist in several different phases, with each phase having its specific reactions and characteristics. Wet deposition processes comprise: (1) in-cloud scavenging (*wash out* – incorporation in cloud droplets, granules, and crystals); (2) below-cloud scavenging (*rain out* – removal by falling rain); and (3) removal by raining clouds (*precipitation scavenging*).

In-cloud scavenging includes nucleation scavenging and collection of a fraction of the remaining aerosol by cloud. Substances scavenged by cloud droplets can be eliminated by wet deposition without rain formation when the cloud comes in contact with the earth's surface. The concentrations of pollutants are up to 10 times higher in cloud drops due to the small size of cloud droplets compared with rain-drops. Cloud deposition may be significant in mountainous regions. Fogs are created by the cooling of air close to the earth's surface and are often formed in polluted areas due to the presence of a high amount of small particles. In such a case, fog is usually enriched with pollutants. During below-cloud scavenging, rainfall may have different chemical compositions depending on the local ambient air pollution. More intensive rainfall washes more effectively and may also be chemically less reactive. Scavenging of air contaminants by raining clouds denotes removal by pre-cipitation (atmospheric water in liquid and solid states). Precipitation depositing on the material surface may infiltrate into the porous structure. Material mass absorbs aqueous solution that is transported inside and causes mechanical stress both by freezing and by hydration and follow-up crystallization of salts. The content of the solute of reactive compounds (such as hydrogen ions from acid compounds) can react with the material and lead to chemical deterioration of the structure. The pre-cipitation can also provide a partial cleansing, as rain removes previously deposited contaminants from exposed surfaces.

Acidic substances that react in air with water and oxidants form acid pollutants and return to the ground as *acid deposition* (in dry form as *dry acid deposition* and in wet form as *acid precipitation*). About half of the acidity in the atmosphere falls back to the ground as dry acid deposition. Damage to materials can result from the dry acid deposition of acid aerosols (such as SO_4^{2-} aerosol) and the dissolution of acid-forming gases on the material surface. The major acid pollutants are oxides of sulphur and nitrogen.

The effects of acid precipitation on materials depend on the type of material and the pH of precipitation. The pH value defines the degree of acidity or alkalinity of water and its solutions and is defined as the negative decadic logarithm of the activity of hydrogen ions H^+, although they are in fact hydroxonium ions H_3O^+ because a proton H^+ never exists in condensed phases and always occurs in the solvated form, i.e. H_3O^+.

Acid solutions formed by scavenging, diffusion, or other processes of involving acidic compounds in the atmosphere can lead to the lowering of the pH of pre-cipitation. The severity of the impact on pH depends on the dissolved substances and the location of the precipitation. In industrialized areas, concentrations of sul-phuric acid and nitric acid in cloud water and in precipitation may be up to 50–100 times greater than their concentrations in areas that are not influenced by emissions of anthropogenic pollutants. Precipitation is normally slightly acidic and has a pH between 5.0 and 5.6 because of natural atmospheric reactions and the dissolution of carbon dioxide – a natural constituent of air. Rain is considered to be acidic when its pH falls below 5.6 (the pH of pure distilled water is 7). For example, as of 2001, the pH of precipitation varied from approximately 4.6 to 4.9 at selected EMEP sites in Europe; in the US, the pH of the acidic rainfall ranged from about 4.3 to 5.3 in 2003.

All forms of acid deposition (especially synergistic effects of dry acid deposition and precipitation) cause serious deterioration if deposited on the surface of materials that are sensitive to acidic compounds. In fact, most materials are susceptible to some degree of acid damage. In exposed areas, the decay can take several forms including roughened surfaces, the removal of material, the loss of carved details from objects or buildings, and the build-up of crusts in sheltered areas. An object's surface may be lost all over or only in certain spots. Especially at high concentrations of contaminants or after prolonged exposures, acid compounds induce numerous undesirable reactions on material surfaces and speed up the weathering process of exposed structures. Those that are most vulnerable are: limestone, marble, carbon-steel, zinc, nickel, copper and copper alloys, and some polymer composites. It is estimated that 10% of chemical weathering of marble and limestone is caused by the wet deposition of hydrogen ion from all acid species. Acid deposition is less harmful for stones with low porosity and resistant compounds containing silicates. Aluminium and stainless steel are more resistant metals.

3.1.2 Effects of gaseous pollutants

The gaseous pollutants that cause the most damage to construction materials include sulphur dioxide, which is the main pollutant with respect to material deterioration, but other pollutants also contribute, especially carbon dioxide, nitrogen oxides ($NO_x = NO + NO_2$), and salts from sea spray. Atmospheric oxidants such as ozone, atomic oxygen, and different free radicals generated during photochemical reactions are responsible for the formation of sulphuric and nitric acids. They are produced via a complex sequence of photochemical reactions and some acid-generating chemical reactions that occur among gaseous components in the air (such as reaction of pollutants with hydroxyl radical) or among dissolved gaseous constituents in atmospheric clouds.

Emissions of sulphur dioxide from natural and anthropogenic sources are the primary sources of sulphuric acid and fine sulphate particles in the air. Atomic oxygen, nitrogen dioxide, hydroxyl radical, and hydrogen peroxide may oxidize sulphur dioxide in the atmosphere. Reactions that convert sulphur dioxide to sulphuric acid, including reactions with hydrogen peroxide in clouds and hydroxyl radical in air, are important. The dominant process is the heterogeneous oxidation of sulphur dioxide and the formation of sulphuric acid in the aqueous phase, which occurs both in the atmosphere and on the material surface. The reaction of sulphur dioxide with the hydroxyl radical can be of importance, in practice, in the formation of sulphuric acid aerosol. Sulphur dioxide deposited on the material surface is absorbed by the surface moisture. The migrating moisture primarily serves as a transport medium for sulphur dioxide that can be translocated internally while being oxidized to sulphates. Sulphur dioxide may react with the material itself in the presence of oxidants and catalysts. Fly ash promotes sulphur conversion since it contains Fe_2O_3 and the oxides of other metals (such as V, Fe, Cu, Cr) that catalyse the oxidation of sulphur dioxide into sulphur trioxide, which thereafter directly reacts with moisture to form sulphuric acid.

Sorption of sulphur dioxide on porous building stones causes physical changes in stones, especially with regard to changes in porosity and water retention. Sulphuric acid aerosol has a corrosive effect on all surfaces, including ceilings and walls of buildings and monuments.

Nitrogen oxides contribute significantly to the total loading of air pollution. In the atmosphere, a major portion of nitric oxide is gradually oxidized to nitrogen dioxide. The removal of nitrogen dioxide from the air occurs via oxidation and hydration to form nitric acid. Oxidation of nitrogen dioxide is performed by hydroxyl radicals and by a heterogeneous reaction involving NO_3 radicals and ozone that occurs at night. Nitric acid is among the main acid components of the atmosphere. The dry deposition of nitric acid is a relatively slow process and its mean time of residence in the atmosphere may reach thousands of hours, during which it is transported over large distances. During the course of wet deposition, nitric acid can be absorbed in cloud and precipitation droplets and fall on the material surface in the form of acid rain.

Nitrogen compounds tend to accelerate the atmospheric attack of materials. Nitric acid decomposes calcium carbonate in inorganic materials by a reaction in which calcium nitrate is formed:

$$CaCO_3 + 2HNO_3 \rightarrow Ca(NO_3)_2 + CO_2 + H_2O.$$

The calcium nitrate that is generated is highly soluble and may be washed off from the surface by rainwater.

Nitrogen dioxide also induces the corrosion of certain metals. In such a case, the deterioration mechanism can be attributed to the oxidation of nitrogen dioxide to nitric acid and its subsequent effect on metals.

Atmospheric contaminants such as chlorine and chlorine compounds when present in the atmosphere can intensify atmospheric corrosion damage. They are released into the atmosphere from natural as well as anthropogenic sources. The corrosive effects of chlorine and hydrogen chloride in the presence of moisture tend to be stronger than those of 'chloride salt' anions due to the acidic character of the former. Apart from enhancing surface electrolyte formation by hygroscopic action, the direct participation of chloride ions in the electrochemical corrosion reactions is likely.

The deterioration of a material may promote synergic effects of different types of gaseous pollutants and particles and/or their combination. Sulphur dioxide is, for instance, more reactive under higher nitrogen dioxide concentrations when increased rates of corrosion occur. This is because nitrogen dioxide oxidizes sulphur dioxide to sulphur trioxide thereby promoting further absorption of sulphur dioxide, and generating sulphur trioxide reacts with moisture to form sulphuric acid.

3.1.3 Effect of atmospheric aerosols
Dry deposition of aerosols (defined as minute liquid and solid substances that are dispersed in air) affects most building structures to a certain extent. Aerosols of

anthropogenic origin are particularly important from the viewpoint of aggressiveness with which they attack a material. Anthropogenic aerosols may contain different inorganic and organic species that include inorganic acids and salts, such as sulphates and nitrates, a number of polar and non-polar organics, as well as metals, which are incorporated in the early stages of particle formation from combustion processes. Sulphate ions are most frequently contained in the atmosphere in the form of acid aerosols. They particularly attack carbonate stones such as limestone, dolomitic sandstone, and cement; however, the acid also affects the stone binders that contain calcium carbonate. Most ferrous metals such as mild steel and galvanized steel are susceptible to corrosion from exposure to acid aerosols. Zinc corrosion induced by sulphuric acid aerosol proceeds at relative humidity greater than 60%. In contrast to this, processed aluminium and aluminium alloys are corrosion resistant because after exposure to a low concentration of acid sulphate particles, a protective aluminium oxide layer is formed. The corrosion of copper and its alloys occurs in the presence of sulphate ions and involves the formation of a protective corrosion layer that is mostly composed of basic copper salts.

3.1.4 Effect of particles

Both primary and secondary atmospheric particles can be of principal importance in the damage caused to materials and building structures. The particles emitted directly (such as sea salt, wind-blown dust, volcanic emissions) are approximately equal to those originating from gas-phase conversions. Gases can become liquid or solid particles on cooling or by chemical reactions in the atmosphere.

The deterioration of materials depends on the chemical composition and properties of the particles. The most aggressive species are salts (chlorides from the seacoast) and several types of fly ash from municipal incinerators. Dust particles that are inert and insoluble in water have practically no impact on materials but may have detrimental mechanical effects through abrasion of the surface of objects or may reduce the aesthetic appeal of a material by soiling. However, such particles can behave as carriers of aggressive chemicals and/or serve as the concentration site for chemically active ions, such as sulphates, nitrates, or chlorides, and may create ideal conditions for oxidation processes on the material surface. All particles are susceptible to electrostatic forces that promote deposition if either the particles or the surface carries an electric charge. The deposition of solid particles from the atmosphere can have a significant effect on the corrosion of some metals because these deposits can stimulate corrosive attack by reducing the critical humidity levels, through hygroscopic action, and by providing anions that stimulate metal dissolution.

3.1.5 Formation of crusts and patinas

The reactions of materials with air contaminants deposited on their surface can lead to the generation of salts that form a layer on their surface. This layer on stones is seen as soiling and is called crust. The corrosion layer formed by the deposition of sulphates and chlorides on copper and copper alloys is called patina.

The properties of the salts that are deposited on the stones vary. Some of the salts dissolve in the superficial moisture layer and may be washed off from the surface by rainfall, whereas other less water-soluble salts, together with dust or soot and other deposited particles (e.g. of biological origin), form a relatively thin, hard, and adhesive layer on the stone surface, whose colour changes from grey to black. These crusts can be observed on the exposed stone surfaces of any structure. The composition, structure, and thickness of the crust differ from the primary material and are dependent on many factors, especially the type and shape of the stone, the shelter against rain and snow, and the type of climate. The composition of each crust, however, is governed by the composition of the particular airborne pollutants in the area. Crust development may be enhanced if the stone is porous or has an irregular surface area.

A significant component of the crust of stones containing carbonates (such as limestone and marble) is calcium sulphate dihydrate (gypsum – $CaSO_4 \cdot 2H_2O$). Gypsum crystals form at the pore opening–air interface, where evaporation is greatest. The gypsum crust readily traps dust, aerosols, and other particles of atmospheric origin; among these, the most abundant are black carbonaceous particles. These black particles, originating from oil and coal combustion, have a large specific area and contain metals, which may catalyse the formation of black crusts. The crust may also comprise calcium carbonate, transported by water in the form of hydrogen carbonate from the primary stone. It behaves as a binder that cements the crust but along with this it also leads to a reduction in the strength of the inner layers of the stone. This process is often aided by bacterial action. Secondary layers and crusts of gypsum may be formed by dissolution and may redeposit in the presence of sulphate.

A typical crust containing gypsum is formed on the surface of marble, which has a high content of limestone and low porosity. A surface that is directly exposed to rainfall is white in colour because the deterioration products that are formed on its surface are continuously washed off (as calcium sulphate dihydrate is approximately a hundredfold more soluble in water than calcium carbonate). On these sides, the surface is rough. An irregular hard crust composed predominantly of crystals of gypsum mixed with black particles covers the black areas found in zones protected from direct rainfall.

Porous sandstones may contain salts that infiltrate from the inside to the surface of the stone. In many types of sandstone, the surface hardening may develop a honeycomb pattern, with crumbling behind the crust and in the cavities. Honeycombs are frequently observed in buildings made of sandstones. Along with salts, the crusts of most sandstone can contain some organic compounds, largely of paraffin type. The presence of this organic matter influences the hydrophobic properties of the sandstone to a certain extent. It may result in the accumulation of salts under the crust and lead to crust splitting. Unlike sandstone, a relatively thin, hard, and cohesive crust originates on the surface of igneous rocks. It usually contains deposited solid particles, including soot. The decay of silicate minerals proceeds very slowly, except tremolite present in some dolomite marbles and black mica in specific marbles or granites.

On copper and copper alloys, basic sulphates and chlorides form a patina. This corrosion layer is not readily dissolved by acids and thus is a good protection for a metal surface. The formation of this layer is controlled by the ability of copper to react with the compounds present and varies with the humidity, the temperature, and the concentration of these pollutants.

3.2 Effects of water

Water damage is considered to be one of the essentially problematic elements in the decay of building materials. Water can deteriorate building structures with which it is in continuous contact and can influence the damage to the material surface by acid deposition reactions. To a certain extent, water may also directly affect the quality of building materials such as concrete during its production. Deterioration is induced by the simultaneous effects of different ions and molecules present in water (including pollutants), by aquatic environmental factors (such as the physical characteristics of water, aquatic macro- and micro-organisms), climatic conditions, resistance of a material to attack, mode of contact between the material and water, and water mass motion.

Water is a good solvent for a number of solid substances, liquids, and gases. Due to its dipolar nature, water is a good solvent for ionic compounds and is particularly suitable for the formation of addition compounds with ionic substances and with substances having a dipolar character. There are very few compounds known that do not dissolve at least to some extent in water.

Water is a medium for the transport of salts in pores, it accelerates chemical reactions that proceed in the material and on its surface, and it may support the growth of micro-organisms (such as algae, lichens). In the areas where the temperature fluctuates around $0°C$, water gives rise to freeze–thaw weathering that especially deteriorates porous building structures. It is estimated that more than three-quarters of all registered building damages have resulted from these effects.

Among the essential inorganic constituents of natural waters, hydrogen carbonate, sulphate, chloride, and nitrate are the main anions and the important cations are calcium, magnesium, sodium, and potassium. In the majority of natural waters, the concentration of these ions decreases in the above-given order. However, in mineralized waters, sodium prevails over calcium, and sulphates and chlorides over hydrogen carbonates. Ammonia and ammonium ions are generally present in low concentrations in these waters. Biological processes significantly influence groundwater composition. Depending on the level of dissolved oxygen, these processes can involve either reduction or oxidation reactions. At greater depths, in the absence of oxygen, anaerobic reactions are observed (particularly reduction of nitrates and sulphates), whereas in the layers with sufficient access to oxygen, aerobic degradation of organic substances takes place. In this manner, aggressive components such as hydrogen sulphide, carbon dioxide, and some inorganic compounds of nitrogen are introduced into water. Seawater is slightly alkaline (the pH varies between 8.0 and 8.3), but in shallow coastal areas carbon dioxide can be accumulated and thus the pH can drop below this value. Seawater contains, on average, 35 g/l of dissolved solids, but this amount is not the same everywhere (the Dead Sea has 280 g/l of dissolved solids).

From the viewpoint of environmental deterioration of materials, the term water usually comprises not only natural waters, but also service (like cooling water, feed water, etc.), industrial waters, and wastewater. There is a huge variation in the composition and content of substances in these waters, depending on the purpose for which the water is used or in which processes originates; thus, various factors determine these parameters. In wastewaters with a predominance of inorganic pollutants, components can be either in dissolved or in suspended forms. Among the dissolved salts, sodium, potassium, calcium, and chloride and sulphate ions can occur most frequently. The composition of sewage water depends on the pollution sources of the contaminants, including kitchens, laundries, bathrooms, and sanitary facilities. The most important components in sewage waters are chlorides, sodium, potassium, phosphates, and inorganic forms of nitrogen – ammonia, nitrides, and nitrates.

Water can enter and cause deterioration of structures in a variety of ways. Rain leakage, condensation, and rising damp are the basic sources of damage caused by water. The moisture and the origin and the course of travel of water determine the extent of damage it can cause through the above-mentioned processes.

Rain leakage influences the dissolution of salts, which may decrease a material's resistance. Rainwater that moves through the pores dissolves the salts that are present inside the material and then deposits them on the surface as the water evaporates. The crystallization of these salts on the surface manifests as efflorescence and that beneath the surface as subflorescence. Efflorescence is the most obvious effect and often only disfigures the face of a building. The effect of the crystallization of the salts beneath the surface is destructive and can accelerate the scaling of the structure and the subsequent deterioration. The conversion of calcium carbonate to calcium sulphate results in a type of efflorescence called 'crystallization spalling'. Rainwater leakage in porous sandstone is an example of the impact of dissolution of the grain cement on stone resistance. After rainwater exposure, the calcareous grain cement in sandstone may dissolve and moves outwards. Along with the dissolution of the grain cement, it may cause loss of coherence of the grain bond.

Rising damp occurs when a building is in direct contact with damp soil or ground water. Moisture from the ground is drawn upwards by capillary action up to a height at which there is a balance between the rate of evaporation and the rate at which it can be drawn upwards by capillary forces. A 'wet line' often associated with efflorescence develops. Groundwater may damage a material that is submerged in water and fluctuations in the level influence the quality of building materials. Rotting of wood is another destructive phenomenon that requires the presence of water to proceed.

3.2.1 Aggressiveness of water

Most frequently, aggressiveness of water against materials is evaluated in connection with the undesirable effect it has on different buildings and metallic pipes. Depending on the content and type of components, aggressive waters may be divided into *waters containing aggressive carbon dioxide, water with a low content of salts*, and *acid and sulphate waters*.

Significant damage to building materials (such as limestone, concrete, mortar, or some metals) results from the exposure to natural constituents of non-polluted rainwater such as carbon dioxide and oxygen. Oxygen has a significant impact on metal corrosion (by the control of the oxygen reduction rate and by the support of protective layers on the surface of the metal). Oxygen enters water via diffusion from the atmosphere and/or is released by photosynthesizing organisms that simultaneously bind carbon dioxide and reduce it to organic compounds. Carbon dioxide is usually not considered a harmful gas but carbonic acid which is formed by the reaction of carbon dioxide with water reacts with inorganic materials containing calcium and magnesium and deteriorates their structures. Aqueous solutions of carbon dioxide can also stimulate serious metal corrosion.

Carbon dioxide is considerably soluble in water; about 2360 mg/l dissolves at 10°C. It is dissolved in water in the molecular form as *free hydrated carbon dioxide* that is denoted by the symbol CO_2 (aq.). Slightly less than 1% reacts with water to form non-dissociated molecules of carbonic acid. Carbon dioxide dissolved in water is called *free carbon dioxide*; the term is used for the sum of the concentrations of free hydrated carbon dioxide and carbonic acid. Dissolved free carbon dioxide can be found in almost all natural waters with pH < 8.3; in running surface waters, carbon dioxide is present only in units of milligrams per litre and does not exceed 30 mg/l of free carbon dioxide; fresh groundwater usually contains several tens of milligrams per litre and mineral water frequently contains more than 1000 mg/l.

The calcium carbonate equilibrium in water is of considerable technical importance in terms of the aggressive effects of water. Calcium carbonate can dissolve only when dissolved carbon dioxide is present, according to the reaction:

$$CaCO_3(s) + CO_2 + H_2O \leftrightarrow Ca^{2+} + 2HCO_3^-.$$

Equilibrium is established between Ca^{2+}, HCO_3^-, and carbon dioxide. Carbon dioxide which is in equilibrium according to this reaction is called *equilibrium carbon dioxide*. If a higher amount of free carbon dioxide is present in water than that corresponding to the carbonate equilibrium, then water dissolves calcium carbonate. When free carbon dioxide is exhausted (e.g. by ventilation) from water, the carbonate equilibrium shifts to the left and calcium carbonate has a tendency to separate from water. The difference between free and equilibrium carbon dioxide is called *excess carbon dioxide*. A part of the total excess carbon dioxide has aggressive effects on some materials. To ensure the non-aggressivity of water and to avoid the corrosive effects of residual carbon dioxide, water must contain a certain minimum amount of hydrogen carbonate.

Aggressive water, which has a low content of salts, influences the process of extraction of soluble constituents of a material. Attack by this type of water depends on the mode of contact between water and the material, the material's resistance to attack, and the water temperature.

Acid aggressive waters contain mineral and organic acids, which dissolve components of building structures, destroy water installations, and may also dissolve the heavy metals contained in them.

Some buildings may deteriorate considerably when exposed to aggressive waters containing dissolved sulphates. The destructive effect of these waters on concrete is more intense due to the formation of strongly hydrated compounds – ettringite, which crystallizes with 30–33 molecules of water, thus considerably increasing its volume and influencing the material via crystallization pressure. Sulphates occur in water predominantly in the form of simple SO_4^{2-} anions. The essential part of sulphates may come from sulphates leached from the soil or from groundwater. In groundwater, the content of sulphates ranges from tens to hundreds of milligrams per litre. They may also be generated by the reaction of carbonates present in building materials with sulphur dioxide. The deterioration effect can be enhanced by the interaction with other dissolved substances that are present in water.

3.2.2 Effect of soluble salts

Salt weathering is probably the most important deterioration process of inorganic building materials (such as stone) in cities, but it may be important in coastal areas as well. There are several sources of soluble salts: they can be present in primary materials, or form during weathering processes, or may also penetrate from the outside, e.g. from groundwater or sea spray in coastal areas. In inland areas, salts of chlorides such as $NaCl$, $CaCl_2$, and $MgCl_2$ are nearly always from road salts, where they are used as winter de-icing agents. Two types of soluble salts may be found in the building materials: salts whose crystals contain crystalline water and salts that can crystallize without water. Salts with crystalline water are less soluble in water and in such cases the formation of crystals from solution is easier. When there is a decrease in the ambient air humidity, they lose crystalline water and form anhydrous salts. The converse effect occurs when humidity increases: salts again bind water. Salts that crystallize without water are very soluble in water. Solutions of these salts diffuse into the pores and infiltrate relatively deep into the material. The concentration of salts increases in the surface layers, e.g. in the sides, where water evaporation is maximum.

The effect of soluble salts depends on various factors, such as the type of material and its porous structure, the nature and quantity of soluble salts, the quantity of penetrating water, and the mode of contamination. In addition, these salts dissolve in water and increase conductivity. The effect is largely attributable to three basic mechanisms: (1) the dissolution of salts and its crystallization in a material's capillaries, channels, and crevices during repeated wetting and drying of the material, resulting in contraction and expansion; (2) the hydration of salts that can exist in more than one hydration state; and (3) the expansion of salts in the pores when salts have higher coefficients of thermal expansion than the material (this factor is less important).

The crystallization of dissolved salts and the associated damage can occur every time a salt-contaminated porous material is wetted and dries out. If the migration of the salt to the surface of the material is faster than the rate of drying, the crystals deposit on top of the external surface and form visible efflorescence, which does not damage the structure. When the migration is slower than the drying rate, the solute crystallizes within the pores at varying depth. The crystals grow and press against the immediate material which can lead to the crumbling and powdering of the structure. The rate of drying after wetting is dependent on conditions such as ambient temperature, relative humidity, and wind velocity. The frequency of contraction and expansion due to salt crystallization may determine the speed at which a material undergoes weathering.

The hydration and dehydration of salts takes place under specific conditions of variation in temperature and humidity when some salts can absorb and release molecules of water. These processes cause the salt to expand and contract and thereby exert pressure on the building material which may cause damage. Cyclic changes of moisture in the presence of hygroscopic salts are particularly dangerous.

If the pores of a material are filled with salts that have higher coefficients of thermal expansion than the material, heating causes the salts to expand more rapidly and exert pressure on the material. This differential thermal expansion may in some cases contribute to defects in the structure.

3.3 Detrimental effects of soil

The deterioration of constructions and equipment buried in the soil can be influenced by the physical characteristics of the soil, its chemical properties, and the micro-organisms living in the soil.

Soil is characterized as a heterogeneous polydisperse system, consisting of solid, liquid, and gaseous phases. The solid phase contains a mineral fraction and an organic fraction. The solid mineral fraction is the dominant part of the soil and it constitutes 95%–99% of the solid phase. Organisms and the debris of organisms form the organic part of the soil. The soil water is the total content of water present in the soil pores in the liquid, solid, and gaseous states. Only that part of the soil water that is able to dissolve electrolytes and gases and to disperse colloids is considered to be the liquid soil component – the soil solution. The composition of the soil solution varies and is dependent on the changes in soil humidity and on the reactions of soil water with the inorganic and organic compounds present in the soil. Among the inorganic substances, the soil solution predominantly contains soluble salts, such as chlorides, nitrates, carbonates, sulphates, and phosphates; as regards organic compounds, different low-molecular weight substances are present. The gaseous phase is the soil air, representing a mixture of gases and vapours filling the vacant spaces in the soil. It is characterized by higher carbon dioxide and water vapour content and lower oxygen content. In the topsoil, carbon dioxide is present in the range of 0.1%–1.0%; in lower layers it may be as high as 5%. Water vapour saturates the soil air and the relative humidity is frequently above 98%. The soil air is also enriched with hydrogen sulphide, hydrogen, sulphur dioxide, methane, and other trace gases, which can have a negative effect on soil air quality. These can stimulate damage in materials.

The deterioration of materials embedded in the soil depends on the composition, the moisture content of the soil, the pH value, the buffering capacity, and the oxidation–reduction potential of the soil solution (the oxidation–reduction potential governs the course of oxidation and reduction reactions). The physical and mechanical characteristics of the soil, such as the porosity and the sorption capacity of the soil, the adhesion and cohesion, the resistance to deformation at a certain soil moisture, and the swelling and carrying capacity should also be taken into account.

With respect to considerably stimulating negative effects, the chemical attack of soil is especially important for buried constructions. The soil's chemical properties may cause the corrosion of embedded steelwork or the decay of embedded timber within walls. The corrosion damage caused by aggressive road salts to transportation infrastructures, including steel bridges, large span-supported structures, parking garages, and pavements, leads to shortening of the service life and the usefulness of structures and thus has serious economic implications.

Soil may contain different harmful substances that have the potential to cause damage to structures. They occur naturally, enter the soil during atmospheric deposition of pollutants, or leak into the soil during storage (such as leakage from underground pipes, tanks, and landfills) and transport of materials. Dry and wet acid deposition of airborne emissions adds hydrogen ions and lowers soil pH; however, acids can also be neutralized by the soil. The extent to which soils can neutralize acid depositions depends on factors such as the type of soil, thickness, weather, and water flow patterns. If the ground is frozen, the neutralization process of the soil cannot function and the acid is not neutralized.

HCO_3^- and NO_3^- anions occur most frequently in soil, the remaining portion corresponds particularly to SO_4^{2-} and Cl^- anions (in non-saline soils); in the soil solution of saline soils, the amounts of sulphate and chloride anions are enhanced. Sulphur is present in the soil in the form of sulphates and sulphides. Predominantly, sulphur dioxide, sulphuric acid, and sulphates enter the soil during dry and wet depositions. The hydrogen ions in sulphuric acid exchange places with the metal ions present in the soil and the calcium, potassium, and magnesium ions are leached or washed out of the topsoil into the lower subsoil. The acid can be immobilized as the soil or vegetation retains the sulphate and nitrate ions (from sulphuric and nitric acids).

The major sources of chlorides in the soil are, in addition to sea spray on the seashore, the road salts used in winter for de-icing and anti-icing. Road salts include mainly potassium, calcium, and magnesium chlorides and magnesium acetate. Sodium chloride is one of the most commonly used de-icers and calcium chloride or magnesium chloride is frequently used in brine solution for anti-icing. The chloride ion often initiates and promotes the deterioration of various concrete components and the corrosion of steel reinforcements. Chlorides penetrate and migrate through the concrete and cause corrosion of steel bars in the concrete. Formation of corrosion products lead to increase in volume and build up stresses which can cause cracks in the concrete. Salts applied to the concrete surface of pavements and roads can increase the frequency of freezing and thawing cycles, thereby increasing the potential for scaling. In addition to the impacts of the chloride anions, the associated cations of sodium, calcium, magnesium, and potassium also have environmental effects. Sodium cation, which is highly soluble in water, can bind to soil particles, break down soil structure, and decreases its permeability.

Porosity is considered the most important soil structural characteristic and it influences the sorption capacity of soil. Porosity is highly variable and depends on the mutual arrangement of soil particles and aggregates, affecting the total volume of pores and their size and shape.

The soil pH considerably affects soil characteristics and is important for most chemical reactions in the soil, for coagulation and peptization of the soil colloids, for ion sorption, and for solubility of many components. Soil acidity is the result of the effects of several factors, particularly the soil composition, the moisture content, and the soil biological processes. The pH of a soil solution usually ranges between 4.0 and 8.5. It is influenced particularly by the presence of dissolved acids and acidic or basic salts from the solid phase and from infiltrated surface or ground water. Intensive biological activity of micro-organisms can cause the soil to become alkaline, resulting in an increase in pH. The buffering capacity of the soil solution is defined as its ability to restrict changes in soil pH. This capacity depends on the presence of weak acids or bases and their salts. A mixture of carbonic acid and calcium bicarbonate or possibly a mixture of phosphoric acid and phosphates is frequently the most effective buffer system in the soil solution.

Cohesion, the ability of the soil to resist external pressure, reduces the disintegration of soil aggregates. It increases with the drying of the soil and is reduced in the presence of increased moisture. The adhesion is manifested by the soil sticking on the surface of the material and depends on the grain size and soil moisture.

Soil saturation by subsurface flow can lead to softening or wash out of the ground below footings. A common problem is the presence of clayey soil beneath the footings. Cracking of buildings originates wherever clayey soils of significant depth are found underlying buildings. Clayey soil shrinks and swells with changes in moisture content, which can cause the buildings to rise or sink.

4 Biodeterioration of materials

Many micro- and macro-organisms cause physical and chemical changes as well as mechanical damage to materials. Bacteria, cyanobacteria, algae, fungi, lichens, insects, animals, and plants have been found to be involved in the deterioration process. Relative humidity and temperature levels determine whether these organisms flourish or exist at all.

Micro-organisms are present in air, water, and soil and may grow and live on material of inorganic and organic origin. Many micro-organisms cause damage through a range of chelating and etching processes. Enzymes secreted by these organisms can be catalysers of chemical reactions that stimulate an attack. Micro-organisms may utilize products of these reactions (e.g. corrosion products) or use certain components from their habitat as nutrients, which are returned to the habitat in the form of excreted metabolites and may affect the habitat. Chemical damage is more important and may arise by excretion and generation of different acids. These acids are capable of chelating metal ions such as calcium and magnesium.

Stone and wood may be damaged when they serve as the substrate for some higher animals (e.g. the larvae of *Plecoptera* or the macroscopic *Ephydatia luviatilis* adhere to a solid base) or are deteriorated by products of metabolic processes. Serious damage may also be induced by animals' excrement. They react with the material surface chemically or serve as nutrients for further attack by micro-organisms.

Occasionally, bushes and even trees are observed on buildings. They can cause mechanical and chemical deterioration of constructions; during growth, plant roots generate surprisingly high amounts of pressure which may damage buildings. The humic acids present in root systems attack the carbonate components of stone and decompose them.

4.1 Effects of micro- and macro-organisms

Viruses, vegetative cells and spores of bacteria, small algal cysts, fungal spores, and spores of lichens are the most common types of airborne micro-organisms that are found near the ground level. The groups of all major categories of micro-organisms can be found in various types of water, whereas in the soil, the predominant groups are bacteria and fungi. Micro-organisms normally found in outdoor air are also present in indoor environment. High microbial concentration is often observed in farm buildings. Extensive growth of some fungi (*Sporotrichum heurmanni*) has been observed on fresh timber in mines, and the fungus was also isolated from the air in mines.

The bacterial chemical action constitutes the major risk for the deterioration of stones. Especially harmful are those bacteria that obtain carbon from carbon dioxide and energy from light or by chemical redox reactions; some are capable of oxidizing inorganic compounds of sulphur and nitrogen to produce sulphuric and nitric acids. Heterotrophic bacteria utilize organic compounds to obtain carbon and produce chelating agents and weak organic acids.

Both ammonia-oxidizing bacteria producing nitric acid and sulphur bacteria oxidizing hydrogen sulphide and elementary sulphur to sulphate influence the pH of the environment where they live and can cause damage to various stone structures. The concentration of sulphuric acid, for example, may increase by approximately 5% in such media.

The majority of bacterial species in the genus *Thiobacillus* use the oxidation of hydrogen sulphide and sulphur as an energy source and convert these forms of sulphur to sulphuric acid. The process takes place only when there is an adequate supply of atmospheric oxygen and hydrogen sulphide (> 2.0 ppm) and high relative humidity. Because a species of bacteria can survive only under specific environmental conditions, the particular species inhabiting the colonies changes with time. For example, one species can grow only when the pH is between 9 and 9.5, but when sulphuric acid, which is an excreted waste product, decreases the pH of the surface below 6.5, they die and another species that can withstand the lower pH takes up residence. The process of successive colonization continues until a species that can survive under extremely low pH conditions takes over. *Thiobacillus thiooxidans* (also known by its common name *Thiobacillus concretivorous*) has been known to grow well in the laboratory when exposed to a solution of sulphuric acid with pH of approximately 0.5. Growth of *T. thiooxidans* and *T. thioparus* has been observed on exposed monuments made of travertine, marble, or sandstone and concrete. The pH of the surface of freshly placed concrete is approximately 11–12.5. This pH does not allow the growth of any bacteria however, the pH of concrete surfaces is slowly lowered over time (by the effect of carbon dioxide and hydrogen sulphide present in the environment) and is reduced to a level that can support the growth of bacteria (pH 9 or 9.5).

Sulphate-reducing bacteria, having the unique ability to convert sulphate and to produce sulphite as metabolite, are well known for the damage they can cause to metals. *Desulfovibrio vulgaris* and *Desulfotomaculum nigrificans* produce hydro-sulphide that induces corrosion of different systems, e.g. oil and gas pipelines, gas distribution system, and sewerage systems. In sewage pipes, the first step in the bacterially mediated process is the generation of a thin layer below the water level. When this layer becomes thick enough to prevent dissolved oxygen from penetrat-ing, an anoxic zone develops within it. In this biofilm, sulphate-reducing bacteria use sulphate ions from the surrounding microenvironment as an oxygen source for the oxidation of organic compounds. The catalysers are the enzymes of the bacteria. The product of oxidation–reduction reactions in these processes is sulphide, which is released.

Micro-organisms may support the corrosion of some metals by using corrosion products during their metabolism. The chemoautotrophic iron bacteria, for instance, oxidize iron (II) compounds. The oxidation takes place under aerobic conditions in which bacteria utilize the liberated energy and the oxidized compounds are precipitated. The genera *Ferrobacillus* and *Thiobacillus* play an important role in the oxidation of iron. *Thiobacillus* are able to oxidize ferrous ions to ferric ions and simultaneously sulphur compounds to sulphates.

Some autotrophic bacteria, which are members of the nitrifying genera *Nitroso-monas* and *Nitrobacter* that employ nitrification as their sole source of energy, facilitate the production of compounds that damage materials. *Nitrosomonas* medi-ate the oxidation of ammonia to nitrite and *Nitrobacter* converts nitrite to nitrate. Nitrification proceeds as follows:

$$3O_2 + 2NH_4{}^+ \rightarrow 2NO_2{}^- + 4H^+ + 2H_2O,$$

and

$$2NO_2{}^- + O_2 \rightarrow 2NO_3{}^-.$$

The typical result of the action of nitrifying bacteria on stone is an increase in its porosity and progressive decay. This type of deterioration is observed predomi-nantly in the ground floor of buildings, in parts where water leaks in and rises by capillary action.

Bacteria may lead to a change in the physical characteristics of wood, e.g. permeability and absorptivity, and cause a loss in strength. Some bacteria such as those belonging to the genus *Cytophaga* are highly specialized; the only substrate they can use as a carbon and energy source is cellulose. They can completely destroy the structure of the cellulose fibres. These bacteria are widely distributed in soil, continental waters, and seawaters. The genus *Clostridium* participates in the anaerobic decomposition of cellulose. Actinomycetes also play an important role in the decomposition of organic material. They constitute a group of bacteria that usually possess a unicellular mycelium with long branching hyphae. They attack resistant substances such as cellulose, hemicellulose, chitin, and keratin and decompose lignin. Actinomycetes occur in both fresh and salt water and in soil.

Although the bacterial processes do induce damaging effects on materials, they can also be useful for the conservation of fine stone artworks. Bacterially activated carbonate mineralization has been proposed as a strategy for the conservation of deteriorated ornamental stone. This conservation method copies natural processes during which many carbonate rocks have been cemented by bacterially induced calcium carbonate precipitation. The selected micro-organism, *Myxococcus xanthus*, is an aerobic soil bacterium. Calcium carbonate precipitation stimulated by *M. xanthus* cements calcite grains by depositing on the walls of the pores and enables effective reinforcement and protection of porous ornamental limestone.

Algae are another group of micro-organisms that grow in the water film on the stone surface and deteriorate it. They have been found on limestone and sandstone of historical objects and listed buildings. Algae may attack stones by exhaled carbon dioxide, which in the presence of water supports the dissolution of the carbonated component of the stone. Some algae living within the stone may contribute to the disaggregation of the stone. Also, mechanical deterioration of stones takes place if algae develop to such an extent that the generated multicellular colonies increase the pressure on the walls of the pores, thereby damaging the stones.

Most fungi are organisms with mycelium as their vegetative structure. The germinating fungal spores put out hyphae, which branch repeatedly to form a ramifying system of hyphae that constitutes the mycelium. Fungi are well adapted to life in the soil, their most common habitat. They are heterotrophs and predominantly aerobic, and therefore usually occur at or near the surface of any organic material. Fungi can destroy the structural integrity of a material mechanically and chemically. Mechanical damage to stone, concrete, and other building materials is caused by the intrusion of the hyphae into the structure and by the contraction and expansion of the mycelium with changes in humidity. Chemical deterioration is more important and may be caused by damaging excretions such as oxalic, tartaric, citric, succinic, and acetic acids. These acids are capable of chelating metal ions. One additional effect of fungal (and also algal) colonization is that it tends to form a film over the surface which blocks the pores. In particular, any moisture that does penetrate will dry out more slowly, the material will stay wetter for longer, and any dissolved salts could penetrate more deeply.

Wood-destroying fungi induce several different types of decay, ranging from the formation of felt mycelia on the wood surface to destruction of wood and rotting, and each species of fungus may attack different parts within the wood. Conditions suitable for fungal growth are temperatures of 24–32°C, air volume of 20% in the wood, and moisture content of 25% or higher. The classes *Basidiomycetes*, *Deuteromycetes*, and *Ascomycetes* are important. The mycelia of *Basidiomycetes* spread on the surface of the wood and bore deep into it. Several groups of this class, such as *Merulius lacrimans*, *Coniphora puteana*, *Poria sinuosa*, *Lentinus lepideus*, and *Gleophyllum sepiarum*, decompose only cellulose and induce the so-called destructive decay of wood. In such cases, a significant loss in the weight and volume of wood may be observed; the wood becomes fragile, gradually changes colour, and turns brown due to the liberation of lignin. This fungal attack frequently causes wood splitting. Other members of the class *Basidiomycetes*, such as *Armillaria mellea*, *Phellinusigniarius*, and *Canoderma applanatum*, induce the decomposition of cellulose as well as lignin; as a consequence of this action, the wood crumbles and becomes lighter. This type of wood decomposition is called white or honeycomb rot. Some genera of the class *Deuteromycetes* decompose only cellulose and cause soft rot that may lead to wood splitting. Some fungi belonging to the class *Ascomycetes*, such as *Ophistoma*, *Aureobasidium pullulans*, and *Cladosporium herbarum,* may cause wood blueing or blackening. These types of fungi usually do not cause the rotting of wood, they only induce colour changes. Aquatic fungi, such as *Monoblepharidales* and *Gonapodya*, are found on submerged twigs in fresh water ponds and springs, and some species of *Pyrenomycetes* (e.g. *Ceriosporopsis*) attack wood submerged in the sea.

Analogous to fungi, lichens also excrete organic acids that attack materials and produce compounds such as salts of salicylic acid and tartaric acid, which also degrade carbonates in an alkaline medium. Lichens have hyphae that can grow through the pores in stones. After absorbing water, they enlarge considerably in volume and affect the walls of pores by applying pressure on them. Lichens are extremely sensitive to gaseous sulphur compounds, which accounts for their rare occurrence in polluted areas.

Mosses often grow on the surface of stones that are covered by humus. They have the ability to absorb large quantities of water and they also produce organic acids.

Pests attack wood in the soil, water, and also in ambient air (outdoor as well as indoor environment). There are different types of insects such as ants, termites, certain species of beetles, and marine insects that frequently cause great damage. Insects cause a loss of wood strength and, in extreme cases, they result in the collapse of wood structures.

Principal sources and literature for further reading

[1]   Air quality criteria for particulate matter (Chapter 9). *Effects on Materials*, Volume II; US EPA's National Center for Environmental Assessment-RTP, EPA Report, 2004.

[2] Benarie, M., *Metallic Corrosion as Functions of Atmospheric Pollutant Concentrations and Rain pH*, BNL 35668, Brookhaven National Laboratory: Upton, NY, 1984.

[3] Betina, V. & Frank, V., Biology of water (Chapter 4), Soil biology (Chapter 8). *Chemistry and Biology of Water, Air and Soil: Environmental Aspects*, ed. J. Tölgyessy, Studies in Environmental Sciences 53, Science, Elsevier: Amsterdam, 1993.

[4] Cody, R.D., Cody, A.M., Spry, P.G. & Gan, G.-L., Concrete deterioration by deicing salts: an experimental study. *Semisesquicentennial Transportation Conference Proceedings*, Center for Transportation, Research and Education, Iowa State University, 1996.

[5] *Conservation of Historic Stone Buildings and Monuments*, Commission on Engineering and Technical Systems, National Academy Press: Washington, DC, 1982.

[6] Moncmanová, A., *Analysis and Determination of Atmospheric Pollution Category at Sites with Most Damaged Rack Construction Buildings*, HI-2, Housing Institute: Bratislava, SR, 1993.

[7] Moncmanová, A. & Lesný, J., Composition and structure of air. *Chemistry and Biology of Water, Air and Soil: Environmental Aspects*, ed. J. Tölgyessy, Studies in Environmental Sciences 53, Science, Elsevier: Amsterdam, 1993.

[8] Pitter, P. & Prousek, J., Chemical composition of water. *Chemistry and Biology of Water, Air and Soil: Environmental Aspects*, ed. J. Tölgyessy, Studies in Environmental Sciences 53, Science, Elsevier: Amsterdam, 1993.

[9] Rodriguez-Navarro, C., Rodriguez-Gallego, M., Ben Chekroun, K. & Gonzalez-Muñoz, M.T., Conservation of ornamental stone by *Myxococcus xanthus*-induced carbonate biomineralization. *Applied and Environmental Microbiology*, **69(4)**, pp. 2182–2193, 2003.

[10] Skalny, J. & Marchand, J. (eds.), *Materials Science of Concrete: Sulfate Attack Mechanisms*, Special Volume, The American Ceramic Society, 1998.

[11] Zelinger, J., Heidingsfed, V., Kotlík, P. & Šimúnková, E., *Chemie v práci konzervátora a restaurátora*, 2nd edn, Academia: Prague, 1987.

CHAPTER 2

Environmental deterioration of metals

P. Novák

Department of Metals and Corrosion Engineering, Faculty of Chemical Technology, Institute of Chemical Technology, Prague, Czech Republic.

1 Introduction

To make metals usable for people, energy must be supplied for their production. This gives rise to an energetically rich product that is not environmentally stable and deteriorates spontaneously, i.e. corrodes. The energy supplied for its production is thus released into the environment. During its corrosion, a metal merges into a more stable state with a low energy content and a less coordinated structure, into corrosion products which in appearance and content resemble their parent raw material – the ore. Therefore, corrosion is sometimes called reverse metallurgy. People's experiences with metals are supported by findings that most of them (except gold and platinum) with their exposure to the environment spontaneously turn into corrosion products. The best-known corrosion product is that of the most common metal, iron – rust. The deterioration of metal objects by turning into rust is mentioned even in the Bible, 'Do not store up for yourselves treasures on earth, where moth and rust destroy, and where thieves break in and steal' (Matthew 6:19).

Since ancient times, corrosion of metals has been understood as an undesirable process, as illustrated in its definition. 'Corrosion is physicochemical interaction between a metal and its environment which results in changes in the properties of the metal, and which may lead to significant impairment of the function of the metal, the environment, or the technical system, of which these form a part' (*Standard ISO 8044 – Corrosion of metals and alloys – Basic terms and definitions*).

No metallic material (with the exception of metallic coatings) is used merely because it is corrosion-proof. The use of a metal depends, in particular, on its primary utility properties, such as mechanical properties (strength, elasticity), excellent electric or heat conductivity, surface look or other properties. To be able to meet its function, a metal should to have sufficient corrosion resistance, which will ensure that the primary utility property shall be usable in a given environment without any serious constraint in the course of the whole service life of a metal product. The period of the required service life may vary to a great extent, from minutes (e.g. rocket nozzles) to centuries (e.g. objects of great historical and artistic value).

Corrosion reduces the primary utility properties; it may even cause a complete loss of function of certain parts (corrosion perforation of a tank wall, cracks in piping, outward deterioration, etc.). Additionally, in most cases corrosion may significantly participate in an unacceptable contamination of the environment. Of course, corrosion is not the only degradation process inherent in the metallic material. Purely physical deterioration, e.g. wear, deformation, formation of cracks and fractures, embrittlement, also takes place. However, given the presence of a corrosive environment physical deterioration co-acts with a chemical deterioration – corrosion. Even if the physical and chemical deterioration processes are insignificant for a given system, the two being complementary together cause the failure of the metal function.

Corrosive environment contains substances that, when in contact with a given metal, cause corrosion. Besides the chemical composition of the environment, temperature and flow rate significantly influence corrosion. The chemical composition is characterized both by the presence of certain components (or their combination) and by the concentration of single components. Corrosion resistance is most significantly affected by the environment pH and by the presence of substances having an oxidation effect.

Since in most cases corrosion is impossible to be suppressed, the term *corrosion intensity allowance* was introduced to indicate a technically acceptable corrosion rate, according to which it is possible for a metallic part to be exposed in a corrosive environment over a long term without unacceptable changes of its utility properties or worsening of other observed parameters (appearance, worsening of heat transfer or electrical conductivity, etc.). A corrosion rate exceeding 1 mm per year is technically acceptable exceptionally, or in cases where the required service life of the part is very short. For corrosion engineering, the values exceeding 1 mm per year are practically uninteresting, as any construction material exposed over a long term should not reach such a state. It can happen only under exceptional conditions, i.e. conditions difficult to anticipate, or because of stray currents or poor choice of material for a given environment. Limits for acceptable corrosion rate depend on the required service life of a part during which it should safely meet its function or on the extent of tolerable contamination of the environment by corrosion products. In the corrosion data surveys, an upper limit of entire resistance is set out to 0.1 mm per year which is valid only for a uniform corrosion and utilization of metals in industrial facilities having their service life between 10 and 20 years, and not for, to give an example, conditions of a long-term atmospheric exposure of metals, for concrete reinforcement, or for metallic materials in a human body (a hip bone substitute, a denture, tooth filling, etc.). Generally, the uniform corrosion rate limit ranges in the interval of approximately five orders, and practically in the interval of three orders (0.1 up to 0.0001 mm per year).

Corrosion deterioration represents a large group of failures caused by the environment affecting particularly the surface of a metal. For the scope of the deterioration

we identify the so-called general corrosion, which progresses over the whole surface exposed to the corrosive environment in a more or less uniform manner, and
the localized corrosion, which progresses more intensively only in some areas of
the exposed surface of metal.

In case acceptable corrosion is accompanied by any form of localized corrosion,
then the technical acceptability is constrained either by the absolute elimination
of localized corrosion or, more often, by the low (acceptable) probability of its
formation.

A still more important viewpoint for the limitation of maximum acceptable corrosion rate is the requirement for the least possible contamination of the environment
by the corrosion products, e.g. in case of human body substitutes, despite pre-
serving of the functional properties of the metal implants, the release of soluble
corrosion products (metal cations) may lead to unpleasant health complications. In
copper coverings without practical reduction of their service life, the atmospher-
ically exposed metal is released in unacceptable amounts into natural waters and
soil. Some of the effective methods of corrosion protection used are restrained by
ecological viewpoints. This is the case with, for instance, chromating, coatings
with lead pigments, and certain types of inhibitors. The coating compositions with
organic solvents are continuously replaced with water-solvent coatings. This is the
main reason why at least equal substitutes are intensively searched for.

The task of corrosion protection is to reduce corrosion losses, to reach technically acceptable corrosion rate and to exclude the origination of localized forms of
corrosion that could endanger a function of the construction material. The selection of construction material and coating constitutes the basic procedures and the
modification of the environment and of electrochemical protection the specialized
procedures. The efficiency of anti-corrosion measures is in a significant manner
influenced by the construction of a metallic product.

In most of the corrosion cases of metals in electrolytes – that is, the environments
where the components are dissociated into ions – the main degradation mechanism
is the electrochemical one, i.e. anodic dissolution. But 'non-electrochemical corrosion' of metals, such as corrosion in gases at higher temperatures, and hydrogen
damage also inhere chemical reactions in their mechanisms.

2 Electrochemical corrosion of metals

2.1 Basic electrochemical terms

The rate of an electrochemical reaction, which proceeds at the electrodes and during
the course of which the oxidation state of the reaction components changes, may
be expressed in terms of the electric current. A reaction during which electrons are
released (e.g. when a metal forms its ions) is called oxidation (anodic process).
Each anodic process is followed by a cathodic process (reduction), during which
the released electrons are consumed. Regardless of whether the process taking
place within the electrochemical cell is spontaneous or forced, oxidation prevails

at the anode while reduction prevails at the cathode. The polarity of the electrodes depends on whether the process is spontaneous or forced.

If the passage of current is forced, the anode is positive (the electrons are consumed), and the cathode is negative (for corrosion, it applies to the electrochemical protection methods and the corrosion resulting from stray currents). If the passage of current is spontaneous, the anode is to the contrary negative (the electrons are released), and the cathode is positive (with respect to the corrosion engineering, it applies to the acceleration of corrosion resulting from galvanic or concentration cells actions).

Compared with a reference electrode, an electrode potential is a measurable value. Standard equilibrium potentials of the electrochemical reactions are upon agreement tabulated against the so-called hydrogen electrode for which the value of its equilibrium potential was stipulated as zero at all temperatures (Table 1). From a macroscopic point of view, no process advances at the equilibrium potential (E_r) that is caused by the equality of the oxidation rate at such potential and the reduction rate (following the same reaction scheme). Thus it deals with dynamic equilibrium (Fig. 1). Oxidation and reduction may theoretically appear in the same area of the electrode surface.

So, as the current passes through the electrode (j), it is necessary for it to surpass the overpotential that is caused by slow transfer of charge from the electrolyte to the electrode (or vice versa), as expressed by $j = j_a + j_c$. Four types of overpotential

Table 1: Standard equilibrium potentials in volts (the electrochemical series at 25°C).

Mg	Al	Ti	Cr	Zn	Fe	Ni	Pb	H	Cu	Ag	Au
−2.36	−1.66	−1.63	−0.91	−0.76	−0.44	−0.25	−0.13	0.00	+0.34	+0.80	+1.45

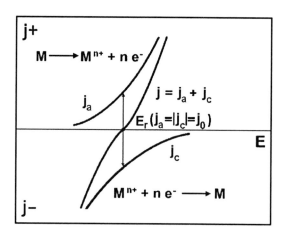

Figure 1: Equilibrium electrode potential.

are distinguished: activation, concentration, reaction and crystallization. Activation (charge transfer) overpotential is caused by a slow process of the charge transfer via the electrode–electrolyte interface. This is applicable throughout the process of electrochemical corrosion but does not necessarily have to be decisive. Concentration overpotential is entailed by a slow transport of electroactive component from the volume phase to the interface and vice versa. As to corrosion, normally the oxygen reduction in the neutral or alkaline environment is concerned. Reaction overpotential is caused by a slow chemical reaction (upon the electrode surface, or in the electrolyte bulk) prior to or after the electrode reaction itself. At corrosion it may deal with weak acids dissociating for formation of reducible H^+, or with the rate of hydrogen atoms re-combination. Crystallization overpotential is significant for electrochemical deposition of metals and is practically not related to corrosion.

2.2 Thermodynamic assumptions of metallic corrosion

The thermodynamic stability of a metal is indicated by the so-called nobility that is represented by the standard equilibrium potential (Table 1). The more positive the metal potential is, the nobler the metal is. With respect to the standard potential value, it is possible to line the metals into the so-called galvanic series. The metals that are considered noble are those whose standard potential is more positive than the stipulated zero potential of the hydrogen electrode. The galvanic series stems from the condition of equilibrium with ions of the relevant metal in solution, but the corrosive environment usually contains very few of these ions and therefore the equilibrium is reached very rarely. The series set up in accordance with the nobility of metals does not take into account the formation of solid corrosion products and is also valid only for aqueous environments without other complex-forming substances.

The formation of insoluble products is reflected in potential–pH diagrams. To begin with, we may consider the potential to be an extent of the oxidation abilities of the environment expressed with the so-called redox potential, which is a measurable value for at-inert (corrosion-proof) electrode. E–pH diagrams define the areas of oxidation abilities and pH, in which either metal (immunity) or its soluble products – cations, alternatively oxoanions (activity-corrosion), or insoluble oxides or hydroxides (passivity) are thermodynamically stable. Borders of these areas are usually determined for very low concentrations of metal ions in a solution and for constant temperature. Lines limiting the stability area of water are set down in the diagrams. Theoretically, water decomposes above this area at the evolution of oxygen, and below this area at the evolution of hydrogen (Fig. 2a).

Areas of conditions corresponding with several types of environment may be marked also for orientation in the potential–pH diagrams (Fig. 2b). The value of potential that is held by a metal constitutes a compromise between the oxidation abilities of the environment and the oxidability of a given metal. Within the area for a given environment, more negative potentials are held by less noble metals in their active state.

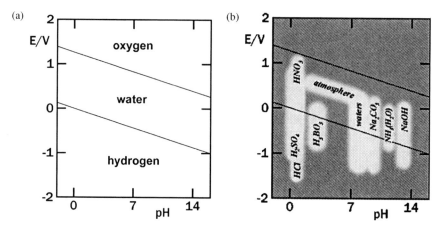

Figure 2: E–pH diagrams: (a) water stability area; (b) areas with conditions corresponding to several types of environment.

The diagrams originated speculatively with regard to the tabulated thermodynamic data, i.e. on the condition of chemical equilibrium between all of the possible species in the metal–water system. Potential–pH diagrams may be exploited for some estimation of metal stability with relation to the oxidation ability and acidity, or alkalinity of the environment; however, it is essential to take into consideration the fact that the diagrams do not convey anything about the reaction rate which is a decisive criterion for the utilization of metals in the corrosion perspective.

In the case of noble metals, immunity overlaps the area of water stability (metals are thermodynamically stable in pure water). The existence of passivity area of a relevant metal (particularly in the area of water stability) enables practical utilization in aqueous solutions. Overlapping of areas of metal stability (immunity + passivity) and areas of water stability is caused by the requirement that water should not spontaneously decompose when the metallic construction material is in contact with the aqueous environment. If we consider the sizes of the immunity area and the passivity area together as the extent of thermodynamic corrosion resistance, then we get to a sequence of metals which corresponds with a general framing experience (the experience that assesses the corrosion resistance in a general way, i.e. regardless of the environment) better than the resistance implied by the electrochemical series. Especially titanium, aluminum and chromium are shifted among the 'noble' metals.

The whole area of water stability is overlapped only by the immunity of gold (Fig. 3a). Should any complex-forming reagent be added to the environment, e.g. cyanides, the border of the immunity area would be shifted to significantly lower potentials, and also gold would, in the presence of the oxidizing agent, actively dissolve (this is used in hydrometallurgical processing). Platinum acquires a large immunity area; nevertheless, in strongly oxidizing environment it is protected by a passive layer.

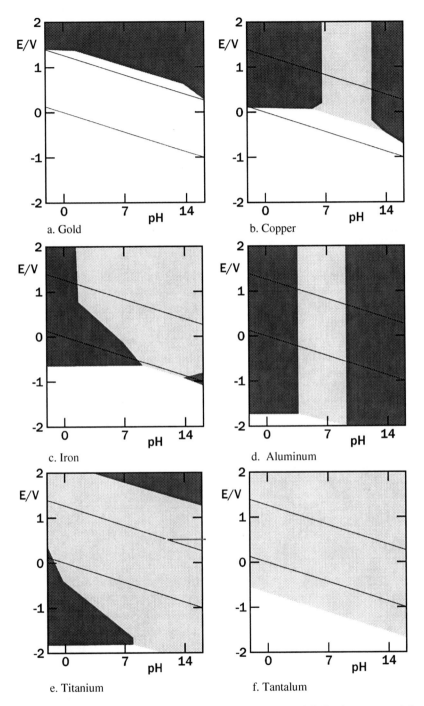

Figure 3: Potential–pH diagrams for different metals at 25°C. Dark area – activity, grey – passivity, white – immunity.

The copper immunity area interferes with the area of water stability only in part; corrosion resistance is normally based on copper oxides stability. Increase of oxidizing ability in the acidic environment is very undesirable for copper; copper does not resist namely in oxidizing acids (Fig. 3b). Silver acts in a similar way as copper, only the immunity area is a little bit larger.

Theoretically, it is possible to use iron, thanks to its passivity in the area of water stability, but it is not possible to use iron in the acidic environment. Both the change of pH and the change of the oxidizing ability of the environment, or potential, lead to passivity (Fig. 3c). Chromium is a non-noble metal with an immunity area lying deep in the region of negative potential values. Similar to iron, the stability of its oxides and hydroxides and the possibility of passivation are exploitable. Zinc and aluminum are amphoteric metals and thus soluble in acidic as well as alkaline environments. Passivity may be reached only by the change of pH, not by the change of oxidizing ability (potential) (Fig. 3d). Titanium and tantalum are very non-noble metals, but thanks to their very stable oxides they are passive almost within entire area of water stability (Fig. 3e and f).

The E–pH diagrams clearly demonstrate that the stability of the two metals most resistant against corrosion, gold and tantalum, is based on totally diverse causes. In case of gold, the entire area of water stability is overlapped by immunity, whereas in case of tantalum by passivity.

A solid corrosion product with a low solubility may not necessarily mean formation of a well-protecting passive layer. In addition, to reach a pH value beyond the neutral area the presence of other ions is essential, and thus the condition for the validity of the diagrams, considering pure aqueous environment, is not fulfilled. In the case of acidic solutions, another anion, which is not considered in the diagrams, is always present; conversely, in alkaline solutions there is a cation. Other ions are also present in neutral solutions. These may significantly influence the areas of metal stability by formation of insoluble products or by formation of soluble complexes, which fundamentally shift the equilibrium conditions. Should another cation or anion change only the oxidizing ability of the environment, then its influence is already present in the potential value at a given pH.

2.3 Corrosion reaction kinetics

In the usual corrosion system not only the species acting on the metal ions equilibrium are electroactive in electrolyte, but several totally diverse oxidation and reduction reactions may also simultaneously proceed at the electrode surface. The total rate of all oxidations has to be equal to the total rate of all reductions, i.e. the sum of all anodic currents has to be equal to the sum of all cathodic currents. The potential spontaneously established by the achieving of the final zero current in a given system is a combination of equilibrium potentials for individually proceeding reactions. Therefore it is referred to as mixed potential. It is not an equilibrium potential, as the equilibrium is impossible to be achieved – a stationary state, at the most. The so-called free corrosion potential is the most frequent case of

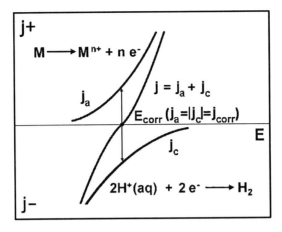

Figure 4: Free corrosion potential for active metal in acidic electrolyte.

mixed potential, which is established while metal anodic oxidation and environment cathodic reduction proceed at the same time (Fig. 4).

The ability of an electrode to be polarized (i.e. to change its potential by current passage) is characterized by polarizability that is closely connected with a value of exchange current density (j_0), which, in case of free corrosion potential, is a value corresponding with corrosion current density (j_{corr}). Out of this value and Faraday laws it is possible to determine the corrosion rate.

2.3.1 Anodic dissolution of metals

At anodic dissolution, metal usually passes into aqueous electrolyte in the form of a hydrated cation. The formation of cation proceeds in several steps. Upon the metal surface, initially the ad-atom generates, which is bound to a water molecule in the other step. The reaction is often facilitated by anions (e.g. OH^- or Cl^-) while an unstable intermediate product is formed. In an aqueous environment, the hydroxide mechanism is always, or at least partly, applied, which may lead to formation of passive layer. With regard to frequent presence of chlorides in the corrosive environment, the so-called chloride mechanism of anodic dissolution of metals is possible and is, for intermediate products, less energy-demanding.

In a number of practical corrosion systems, the chloride and hydroxyl anions mutually compete, as the hydroxide mechanism for, e.g. iron and its alloys, enables to form an oxide-type passivity, whereas the chloride mechanism does not. In these cases we may distinguish between the conditions for passivation and activation of a given metal merely upon the knowledge of the ratio of chloride and hydroxyl ions concentrations in the environment, and of the boundary condition specific for the given metal material. The chloride mechanism is an example of catalytic anions effect. It resembles the sulfate mechanism, which applies to iron under atmospheric conditions.

2.3.2 Cathodic reactions

The process of anodic oxidation of metal is conditioned by the consumption of electrons released during its ionization by cathodic reactions. These are the electrode reactions in the course of which a species of the environment is reduced. Such cathodic reactions have to proceed at the same potential as the anodic reaction of the metal dissolution. In aqueous electrolytes, cathodic reductions of oxygen or hydrogen ions and/or water (evolution of hydrogen) are the most usual one.

Hydrogen evolution may be described with the following two equations:

$$2H^+ + 2e^- \rightarrow H_2 \text{ (in acidic electrolyte)}$$

$$2H_2O + 2e^- \rightarrow H_2 + 2OH^- \text{ (in neutral or alkaline electrolyte).}$$

The first one is applied mainly in the acidic environment with sufficient amount of hydrogen ions; the second one is applied mainly in neutral and alkaline environment (especially at cathodic polarization). Both of these reactions may be followed by corrosion of non-noble metals. The rate of hydrogen evolution depends in a great extent on the electrode material. The rate with which the environment is able to reduce itself upon the metal surface may be decisive for the metal corrosion rate. Potential dependence of the cathodic reaction rate and the value of its exchange current density are essential in this case (Table 2).

The most usual cathodic reaction in electrolytes which are in a contact with the air atmosphere is oxygen reduction which proceeds in accordance with the following equations:

$$\tfrac{1}{2}O_2 + 2H^+ + 2e^- \rightarrow H_2O \text{ (in acidic electrolyte)}$$

$$\tfrac{1}{2}O_2 + H_2O + 2e^- \rightarrow 2OH^- \text{ (in neutral or alkaline electrolyte).}$$

With respect to low oxygen solubility in the aqueous environment, the corrosion rate at corrosion in active state may be limited by slow oxygen transport. Oxygen reduction may proceed even in the area of water stability and thus may follow the corrosion of noble metals (gold only in presence of complexing substances). Regarding the fact that the most common corrosion system consists of a combination of iron and neutral aqueous solution, oxygen reduction supports the transformation of the greatest volume of metallic material into corrosion products.

2.3.3 Heterogeneity in corrosion systems

Anodic and even cathodic reaction may theoretically proceed in the same area of metal surface, and as to the course of corrosion process, geometrically separated

Table 2: Order values of exchange current density (j_0) (A/m^2) for hydrogen evolution on different metals in acidic electrolyte at 20°C.

Metal	Pt	Fe	Ni	Cu	Sn	Al	Zn	Pb
j_0 (A/m^2)	10^1	10^{-2}	10^{-2}	10^{-3}	10^{-4}	10^{-6}	10^{-6}	10^{-9}

anodic and cathodic areas are not necessary. However, there are heterogeneities in real corrosion systems that facilitate the process of anodic or cathodic reaction. Therefore, the reduction proceeds in a simpler manner in areas with lower overpotential and anodic reactions where metal is more easily ionizable.

Heterogeneities are inherent in the structure of most metal materials. These differences lead to activities of the so-called microcells which are given by structural arrangement inhomogeneity (crystal planes, lattice defects), or chemical inhomogeneity (gradient of chemical composition of a phase, various phases, precipitates, inclusions – electron-conductive or chemically unstable in environment) or inhomogeneity of the internal mechanical stress.

In addition to microcells activities, cells with anodic and cathodic areas at a distance of more than mere tens of μm frequently form within the corrosion system. Such cells are designated as macrocells and the distance to which the metal potential is influenced by their activities may reach even hundreds of meters (e.g. in case of lined buried facilities – piping and cables).

Connections of two differently noble metals having dissimilar anodic behavior may serve as a classical example. The metal with more negative free corrosion potential turns into anode, and the metal with more positive potential becomes cathode. Activities of such galvanic cells lead to the acceleration of the anodic metal corrosion, and frequently, as a result of geometrical arrangement, to localized attack.

The heterogeneity of the environment that is in contact with the metal surface often leads to the formation of cells that are called concentration cells. The most common example of such cells is the one with differential aeration, when the cell is activated by different accessibility of a certain part of the surface in neutral aqueous environment for atmospheric oxygen. The prevailing anodic reaction upon the surface in lesser contact with oxygen leads to the acidification of the adjacent electrolyte as a result of metal ions hydrolysis. Electrolyte aggressiveness at the anode increases (also as a result of migration of frequently present chlorides), and electrolyte aggressiveness at the cathode decreases. The final consequence is the passivation of the surface with better access to oxygen and increase of corrosion of the surface where access to oxygen is limited. The original heterogeneity in access to oxygen has therefore turned into heterogeneity of composition (pH, Cl^-) initiated by the corrosion cell activity. At construction crevices, in cracks, semi-closed pits, at level meniscus, under sealings, coatings, deposits and corrosion products, the same mechanism leads to the formation of what are known as occluded solutions. Likewise, the microbial activity upon the metal surface leads to formation of occluded solutions and concentration cells. Formation of occluded solutions follows the localized forms of attack, such as pitting and crevice corrosion, stress corrosion cracking, corrosion fatigue, etc.

Heterogeneity in the composition of the environment does not have to be related to access to oxygen but may have connection with concentration differences of other species of the environment that affect the corrosion process. Likewise, heterogeneity in physical parameters at the metal surface (temperature, heat flow, flow rate) leads to formation of cells.

2.3.4 Passivity of metals

Should the suppression of corrosion reaction on a metal be caused by a protective layer generated spontaneously upon the surface (referred to as passive layer), we call it a passive metal. Various barrier effects of non-oxide corrosion products are involved in constituting passivity. The oxidation type of passivity is caused by a very thin, non-porous layer of oxides-hydroxides on the phase interface metal–electrolyte. Salt passivity occurs almost in every metal under conditions where the solubility product of forming salts is exceeded upon their surface.

Most frequently, upon the passive metal surface, there is a very thin layer of hydrated oxides which is of amorphous, almost microcrystalline character, usually is electron-conductive (very little conductive at aluminum, titanium, and tantalum), and whose composition is given by the basis metal (alloy). The passive layer becomes continuously enriched with some elements, such as chromium and silicon in stainless steels. It is anticipated that the protective effect is given by the function of the layer as bipolar membrane, which prevents penetration of anions from the electrolyte (outer part) and penetration of metal cations into the electrolyte (inner part). The penetration of OH^- ions is essential for the formation of the layer from the metal–layer interface ensured by the splitting of molecules of water by electric field within the layer.

In a number of electrolytes, passivable metals exhibit dependence of corrosion rate on environment oxidation ability corresponding with the curve from Fig. 5. There are four areas with different corrosion resistance. In the area of immunity, metals are thermodynamically stable. In the area of activity, the metal oxidation rate, expressed by corrosion current, is exponentially dependent on the oxidation ability of the environment expressed by a potential. Increase of oxidation ability of the environment leads to increase of corrosion rate. In active passivable metals, after oxidation under given conditions has reached a certain level there is a sudden drop in the corrosion rate, which in a relatively broad interval of conditions sustains low value – the metal is passive.

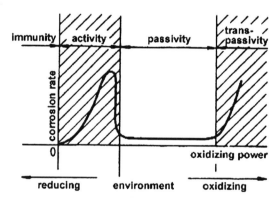

Figure 5: Dependence of passivable metal corrosion rate on oxidation ability of environment.

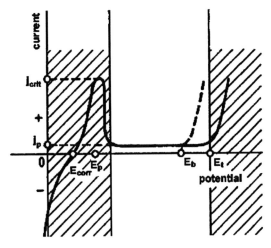

Figure 6: Dependence of current on the potential for passivable metal.

Any further increase in the oxidation ability of the environment results in the failure of the protective function of the passive layer, and the corrosion rate significantly increases. Where the environment does not contain ions locally damaging the passive layer, we talk about transpassivity in which metal corrodes practically uniformly all over the surface. If there are any aggressive ions in the environment (e.g. chlorides for steels or aluminum), it causes local breakdown of passive layer and localized attack of metal.

Qualitative dependence, expressed in Fig. 5 in general coordinates corrosion rate–oxidation ability, may be expressed in coordinates current–potential (Fig. 6) where current is an expression of the electrochemical reactions rate upon the metal surface, and potential is a formulation of oxidation conditions upon its surface. This experimentally observable dependence is a sum of partial currents of all electrochemical reactions proceeding upon the surface. In the whole potential range, the polarizing curve does not express merely the corrosion rate of metal; it corresponds with the observed current at potentials at which neither reduction nor oxidation of the solution species occur. The basic characteristic current values in dependences are passivation current density (j_{crit}) – current necessary for passivation of metal surface unit (10–10^4 A/m^2) and passive current density j_p which usually corresponds with the corrosion rate of passive metal. Passive current largely depends on time and partially on potential and reaches values of 10^{-4}–10^{-1} A/m^2.

Passivation potential E_p is the potential the exceeding of which results in passivation. Potential, at which metal may be considered as persistently passive, is often somewhat more positive, and thus the passivation potential cannot be considered as an unambiguous border of usable passivity on the negative side. Transpassive potential E_t limits the passivity on the positive potentials side from transpassivity. In case a solution contains aggressive ions causing local breakdown of passive layer, the stable passivity is limited by breakdown potential (E_b) on the side of positive values. The range between passivation and transpassive or breakdown potential is a passive area in which corrosion protection is applied by passivation. This area is broad, of the order of hundreds of millivolts; for titanium, tantalum and aluminum, it is of the order of tens of millivolts.

2.3.4.1 Factors affecting passivability and corrosion resistance in passive state
Passivability is considered to be a range of adversity for a metal's transfer in a given environment from active to passive state. It is identified by values of passivation potential and passivation current density. Corrosion resistance in passivity and the stability of passive state are characterized by a value of passive current density and width of the passive range, i.e. potential difference between passivation potential and transpassive or breakdown potential.

Technically significant passivable metals are iron, nickel, chromium, aluminum, copper, titanium, tantalum and their alloys, such as stainless steels and nickel alloys. The composition of a metal is the most distinctive factor affecting its passivability, even the corrosion rate in the potential area of passivity.

Passivability of metal is simplified by alloying elements that decrease the passivation current density and shift the passivation potential to negative direction. Stability of passive state is supported by metals that decrease passive current and broaden the passive range.

Passivability of iron alloys is most significantly influenced by chromium which, added in amount exceeding 12 wt.%, considerably simplifies passivation. The mechanism of chromium effect may be explained with a shift of passivation potential to negative direction that enables easy, spontaneous passivation. Neither decrease of the rate of a partial anodic reaction should be omitted. Iron alloys containing more than 12 wt.% of chromium (stainless steels) have corrosion resistance just from easy passivability and resistance in passive state. Nickel in iron alloys decreases passivation current and the corrosion rate in active state.

In addition to composition, another important factor is the metal structure, which may be affected by the composition as well as heat or mechanical treatment of the material. Formation of structural non-homogeneity of metal constricts the passive area and increases characteristic currents. This usually results in occurrence of localized forms of attack, such as intergranular and pitting corrosion.

The environment affects passivability of a metal and its resistance in passive area mainly by composition and temperature, and alternatively by motion. Metals passivate not only in aqueous solutions but also in environments with distinctively low water content, such as certain organic solutions or in waterless solutions as a melt with ionic conductivity.

The influence of chemical composition is given by the content of species with activating or passivating effects. Unambiguous distinction of solution species regardless of the protected metal is impossible, as the components with activating effect on one metal may have passivating effect on another.

Substances with oxidation effects normally come under the category of passivators. Chloride, fluoride and sulfide anions usually have activating effects.

Several different states may be a result of the oxidation effect of the environment on a metal which is passivable under given conditions. Where the oxidation ability is low for passivation as well as for the sustaining of the metal in the area of passivity, the metal corrodes in activity. Should the oxidation ability of the environment be sufficient to sustain the metal in the area of passivity but insufficient to passivate the active metal, then the passive state is unstable and any local failure of the passive layer leads to overall spontaneous activation. Should the oxidation ability of the environment be sufficient for spontaneous passivation and the passive state is stable, then the passivity is a basis for the corrosion resistance of the metal. In case the oxidation ability of the environment is too high, corrosion proceeds in the area of transpassivity, or the pitting corrosion occurs above breakdown potential, then the passivable metal does not resist any longer.

2.3.5 Biologically influenced corrosion

In some cases, the corrosion attack of construction materials is impossible to be explained merely on the basis of abiotic corrosion mechanisms. In most cases microorganisms participate in the stimulation of the corrosion process. Various types of organisms are present in all natural environments. In aqueous environments, the microorganisms incline to attach to the surface of solid substances and grow thereupon. It results in formation of more or less uniform layer of biofilm. The activity of microorganisms in the biofilm upon the surface of metals causes a change of chemical composition of the environment and physical conditions on the interface metal–environment.

In the first case, the microorganisms stimulate the corrosion process by using the environment components or corrosion products during their metabolism. The ability of sulfates reducing bacteria to exploit hydrogen and the ability of iron bacteria to oxidize Fe^{2+} to Fe^{3+}, and thus permanently support the corrosion process, may serve as examples. Microorganisms may also evolve aggressive substances for metals (evolution of hydrogen sulfide by sulfates reducing bacteria, production of sulfuric acid by sulfates oxidizing bacteria, etc.). Only the formation of biofilm enabling the formation of concentration corrosion cells shall be considered as a physical change on the interface metal–environment. These two changes that mainly influence the corrosion process simultaneously do not lead to the origination of a new type of corrosion attack but affect the initiation or the rate of already known corrosion processes.

The origination of heterogeneities of a physical and chemical nature on the metal–environment interface following the formation of biofilm is characterized by changes of the environment (pH, oxidation ability, temperature, rate and character of flow, concentration of certain components of the environment, etc.). In this respect, the value of a certain parameter on the surface of the metal under the biofilm may totally differ from the value outside the biofilm, and conditions appropriate for a function of concentration cells may arise. It may result in change of corrosion attack arrangement (e.g. from uniform to localized), or in the initiation of the corrosion attack in conditions under which the attack would not occur if that biofilm did not exist (under abiotic conditions).

3 Types of corrosion

3.1 Uniform corrosion

Uniform corrosion is a general corrosion that proceeds at an almost identical rate upon the whole surface. It occurs in case when the complete surface is either in active or passive state. In uniform corrosion in active state, those microcells significantly apply that are mainly uniformly distributed within the surface. As a result of metal corrosion, the localization of anodic and cathodic areas continuously changes. This localized attack on a microscopic scale leads more or less to uniform loss of the metal upon the surface.

3.2 Galvanic and concentration cells (macrocells)

The action of macrocells (see Section 3.3) intensifies corrosion upon the surface of the metal which is anodic under given conditions (it has more negative corrosion potential) (Fig. 7) and conversely, corrosion of the material which is cathodic is reduced by connection. The action of the cell is always developed by heterogeneity of composition or other conditions which acts even during the acceleration of uniform corrosion (microcells) and is related to a range of other types of localized corrosion.

The acceleration of the corrosion rate upon the metal surface caused by the cells activity depends on many factors. In addition to the difference of free corrosion potentials in a given environment and the polarizability of connected metal surfaces, electrolyte conductivity, geometrical arrangement and the ratio of surface areas also factors in this process.

For the cell activity, electric connection of at least two materials is essential, while the cathodic material does not have to be metallic. As the conductivity of

Figure 7: Deterioration of iron bolt caused by contact with copper sheet.

electron-conductive materials, metals in particular, is high, the resistance of electrodes and their connection should not be constraining for the function of the cell. The most important is the electrolyte resistance, which is remarkably higher due to low conductivity of the corrosion environment as well as frequently due to geometrical arrangement. For instance, cells in atmospheric conditions may act in a very thin layer only, and therefore their reach is small and the effect is localized. Vice versa, in the volume of conductive electrolyte, corrosion of anodic metal is intensively influenced by the effect of the cell, but the effect is not so localized in the area adjacent to the connection (though the total effect is the greatest in this place).

In conductive environments, the extent to which cells are affected by corrosion also depends to a large extent on the ratio of the areas of anodic to cathodic materials. When a small anode is connected to a large cathode, the intensification of corrosion reaches its maximum. Conversely, when the anodic surface (compared with the cathodic surface) is large, the extent of corrosion exhibited is small. Due to the high resistance of a thin layer of electrolyte, the ratio of the areas of connected surfaces is not considered to be a distinctive factor under atmospheric conditions.

Protection against the effects of macrocells consists namely in the selection of an appropriate combination of materials. It is possible to connect materials with considerably close, free corrosion potentials, or materials the connection of which does not change the corrosion potential outside of the passive area. Corrosion impacts of the macrocells may be limited by decrease in corrosion aggressiveness of the environment. In some cases it is possible to eliminate the cells activities by electrical insulation (e.g. insulating flange), cathodic protection or extension of the closest distance of the surfaces in electrical contact.

3.3 Crevice corrosion

Crevice corrosion is a localized form of attack which is related to narrow crevices or slots between metal surface and other surface (metallic or non-metallic). One of the dimensions at the slot orifice is usually smaller than $10\,\mu m$. Though it enables the electrolyte inside the crevice to ionically communicate with the electrolyte volume outside, the dimensions impede convection and limit diffusion. Such situation occurs in case of construction crevices, thread connections, pores of welds, locations where a weld is not continuous, at the edge or riveted joint, under sealings (Fig. 8), under deposits, under corrosion products or under disbonded coatings. The crevice effect follows a range of other forms of corrosion, such as stress corrosion cracking, corrosion fatigue, intergranular and pitting corrosion. In the first phase of crevice corrosion, oxygen is relatively quickly consumed from a very small inner electrolyte volume. In the second phase, a slow corrosion of the inner passive surface continues with the cathodic reaction outside the crevice, and the excess of positive charge in the inner electrolyte is compensated by migration of anions (usually chlorides) from the outside electrolyte to the inside of the crevice. In the third phase, the increase of chloride concentration, followed by hydrolysis, continues.

Figure 8: Crevice corrosion of stainless steel under gasket.

Inside the crevice it leads to a formation of an occluded solution which leads to activation of the originally passive surface.

An example of this is the formation of crevice corrosion of stainless steel in aerated water with a content of chlorides that still enables a long-term stability of the passive state. Crevice corrosion of stainless steel in water may be anticipated on the basis of various models which use mathematical description of migration, diffusion, chemical equilibrium and even empirically assessed conditions for activation of various types of stainless steels.

Protection against crevice corrosion consists of construction modifications which impede the formation of construction crevices, discontinuous weld seams, preventing construction connections from forming crevices and restricting the formation of deposits. It may also be restrained by the elimination of porous sealing materials and materials releasing aggressive components. The problem of crevice corrosion may obviously be solved by selection of appropriate material with greater corrosion resistance (which resists even to occluded solutions), or by modification of the environment.

3.4 Pitting corrosion

Pitting corrosion is localized corrosion process in which deep pits are generated upon the metallic surface and the surrounding surface remains without observable attack. This form of attack is generated on a number of passivable metals; however, it is typical for stainless steels (Fig. 9) and aluminum. In the incubation phase, pitting corrosion of stainless steels is normally generated on the basis of competition between hydroxide and chloride ions at the surface. The chloride ions attack the

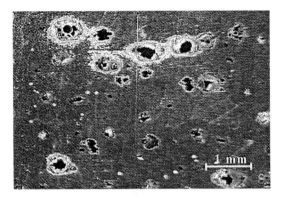

Figure 9: Pitting corrosion of stainless steel sheet.

passive layer, the hydroxide ions form it. The initiation mechanism consists in attacking the metal in areas with lower protective properties of the passive layer (inclusions, grain boundaries) in the presence of various aggressive ions in solution. In the area of passive layer failure, a pit is generated, inside which the concentration of aggressive ions increases by migration, and the value of pH decreases by the hydrolysis of corrosion products. Localization of the attack is caused by actively corroding inside of the pit, which serves as a sacrificial anode for the remaining passive surface.

The generation and propagation of pitting corrosion are enhanced by increased concentration of an aggressive ion (usually chlorides), by presence of oxidizing substances, increased temperature and low pH value.

Protection against pitting corrosion is based on selection of resistant metallic material. Improved resistance of stainless steels may be achieved, e.g. by alloying with molybdenum. The decrease in the probability of pitting corrosion development is supported by the decrease in the environment oxidation ability, the presence of certain ions with an inhibitive ability (e.g. NO_3^-, SO_4^{2-}) and the increase of pH value. Pitting corrosion is supported also by roughly machined surface, occurrence of scales and inclusions (sulfides) in the metal structure. Motion and reduced temperature of the environment help to eliminate the pitting corrosion. It is possible to exploit the electrochemical protection, mainly cathodic, which decreases oxidation conditions upon the surface by polarization. Aluminum and its alloys suffer from pitting corrosion; beside chlorides, aluminum is also attacked by very low concentrations of noble metal ions (Cu, Hg). Pitting corrosion occurs even in copper in certain types of water.

3.5 Intergranular corrosion

Intergranular corrosion is a form of localized corrosion attack which is manifested in stainless steels after heat treatment, during which, on the grain boundaries, it leads to

Figure 10: Intergranular corrosion of stainless steel (specimen after bending).

generation of areas depleted of chromium as a result of formation of carbides rich in chromium, or of other phases with different electrochemical properties. The metallic material corroding intergranularly loses its mechanical strength without any visible surface change. It is because under certain conditions, it preferentially corrodes in a narrow band at the grain boundaries, which lose then their mutual cohesion (Fig. 10). The area at the grain boundary becomes inclined to be attacked by corrosion in an aggressive environment, mostly as a result of decrease of chromium under the limit of 12%, which enables easy passivability of alloy. The decrease of the chromium content at the grain boundary is caused by precipitation of carbides with high content of chromium ($Cr_{23}C_6$, M_6C) at heat treatment in a critical temperature area. The inhomogeneity generated by heat treatment is reversible – it is possible to remove this sensitization by annealing at the temperature of approximately 1050°C, followed by immediate cooling.

The critical area of heat treatment (sensitization) is allocated by the time and temperature of isothermal annealing. Mainly the content of carbon affects the sensitization of austenitic stainless steels – the less carbon is contained in steel, the less sensitive the material is. Intergranular corrosion is fastest under conditions when the depleted part of metal corrodes in active state while the remaining surface is passive.

Intergranular corrosion of 'active boundary–passive grain' type may be prevented by heat treatment beyond the area of critical temperatures, or where applicable, may be protected by the so-called stabilization by adding titanium or niobium which has greater affinity to carbon than chromium is. The presence of carbon in austenitic stainless steels is not essential for absolute majority of their applications due to their mechanical properties; therefore utilization of low carbon steels (<0.03 wt.%) constitutes the best solution for elimination of the danger of austenitic stainless steels intergranular corrosion after welding.

Intergranular corrosion attack is not solely related to austenitic steels but also concerns all other stainless steels and nickel alloys – in addition also alloys of aluminum, copper, zinc, silicon and lead. The so-called exfoliation may serve as an example of intergranular corrosion of aluminum alloys (Fig. 11).

Figure 11: Exfoliation of aluminum alloy beam from aircraft.

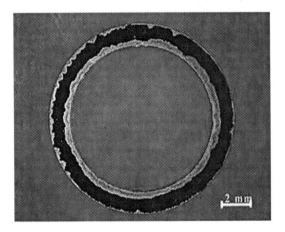

Figure 12: Brass tube dezincification in potable water.

3.6 Selective corrosion

In selective corrosion, corrosion processes cause removal of one component of alloy. A typical example of this form of corrosion is dezincification of brass, when part of the original material, alloy of zinc and copper, turns to spongy copper. It happens either in the whole layer at the surface (Fig. 12), or locally. The spongy copper has no strength, and it continuously causes wall perforation. Dezincification plays its role in the formation of corrosion cracking of brass. All brasses having a Zn content higher than 15 wt.% incline to dezincification. The mechanism of dezincification lies in dissolution of Cu and Zn, and Cu subsequently re-deposits. The dezincification often happens in waters containing chlorides and is a frequent cause of failing of the brass fittings in water circuits. Dezincification is significant in waters with high content of oxygen and carbon dioxide, at low or no flow.

Acidic waters with low salinity usually lead to uniform dezincification at normal temperature; neutral waters with high salinity lead to local dezincification especially at elevated temperatures.

Where a problem with dezincification may occur, it is recommended to use red brass (<15% Zn) (mostly they are inappropriate due to their bad mechanical and processing properties) to limit the oxidation ability of environment by decreasing the oxygen content to increase flow, apply cathodic protection, alloy brasses (α) with arsenic, tin, and phosphorus – the so-called admiral brass.

Other cases of selective corrosion may be observed in aluminum bronzes (dealuminification), silicon bronzes (desiliconification), tin bronzes (detinification), gold alloys (decopperification) and dental amalgams (selective removal of tin).

Graphitic corrosion of gray cast iron with continuous graphite network is another example of selective corrosion. The graphitic corrosion appears in water and in soil. Graphite is continuous in the gray cast iron structure and forms a distinctively noble phase compared with iron. That leads to galvanic cell and selective corrosion of metal matrix. The original gray cast iron turns to non-metallic material in which the iron is completely transformed into corrosion products and has practically no strength.

3.7 Environmentally induced cracking

Straining metal below the yield strength or fatigue strength may, even in environment with low corrosion aggressiveness, result in cracking. The tensile component of stress is always remarkable; the compressive component does not cause the attack. From a technical point of view, this is a very important degradation mechanism as it causes sudden failure of metallic materials, often with high strength and corrosion resistance. Cracks develop, while any other more significant damage of the material surface does not occur.

Various forms of cracking induced by the environment have the same result for users, while the causes and conditions of its development are different. Individual forms have several features in common; there is no exact border between them. In aqueous solutions, cracking developed by the environment may be divided into three types: stress corrosion cracking, corrosion fatigue and hydrogen-induced or -assisted cracking.

3.7.1 Stress corrosion cracking

Corrosion cracking occurs when static tensile stress affects metallic material exposed in specific environment (at certain temperatures and concentrations). In corrosion cracking, cracks propagate in the metal structure, either at grain boundaries – intergranularly (Fig. 13) or via grains – transgranularly (Fig. 14). The mechanism of stress corrosion cracking of a number of technically important materials is explained by active dissolution in the crack front which simultaneously acts as a stress concentrator. The outer mechanical stress is not always necessary for the development of corrosion cracking, but the residual stresses are sufficient, e.g. after

Figure 13: Intergranular stress corrosion cracking.

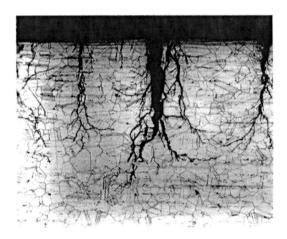

Figure 14: Transgranular stress corrosion cracking.

cold working or at weld seams. The more homogenous the material is, the less susceptible it is (pure metals are the most resistant).

Each irregularity of the surface may cause a localization of stress – tensile stress is greater in this area. Therefore, the formation of a crack is frequently related to the area of crevice or pitting corrosion – the crack may propagate more easily out of a pit or from a crevice; a specific, more aggressive environment develops therein – occluded solution. A typical example of corrosion cracking is the so-called caustic embrittlement. It would occur in boilers made from carbon steel for which water was alkalized to eliminate uniform corrosion (pH above 9). In areas of riveted joints, hydroxide got concentrated by overheating, and the tensile-stressed rivets cracked.

Another example is the so-called season cracking of brass. Brass cartridges with great inner stress (in the area of a pressed-in projectile) in humid atmosphere, in the presence of ammonia vapors (decomposition of organic waste), crack, and thus the ammunition is depreciated. Up to now, this form of failure of brass may be observable, and the main cause is the inner stress originated in cold working. The corrosion cracking of carbon steel in soil environment (in the presence of CO_2 (aq.)/HCO_3^- or HCO_3^-/CO_3^{2-}) which relates to gas pipelines behind the compression stations, where temperature is elevated, is also significant. Another considerable problem is also corrosion cracking of stainless steels which crack in chloride environment. The structure of austenitic chromium-nickel stainless steels is susceptible; therefore, it is more appropriate to use duplex steels (growing of the crack is stopped in ferritic grains) or ferritic steels.

Protection against corrosion cracking is based on the decrease of tensile stress in material and even in construction, on elimination of specific components from the environment, or on transformation of metallic material which under given conditions does not suffer from corrosion cracking. Electrochemical protection and addition of inhibitor may be effective as well.

3.7.2 Corrosion fatigue

Corrosion fatigue is a form of attack which requires co-operation of a corrosive environment and cyclic mechanical stress with a tensile factor. While in cyclic stress, the metal integrity is violated by development of cracks, even if stressed below the yield strength and even without any corrosive environment. In corrosion fatigue, the crack is transcrystalline and is propagated discontinuously that results in development of striations. Frequency of stress greatly influences the attack: the lower it is, the greater the change of the crack length per cycle is. The frequency of 10^{-6} Hz (approx. once a week) is sufficient for the process of corrosion fatigue. In fast opening and closing cracks (at high frequencies), the corrosion attack is not enough to apply, and the mechanical effect prevails. In some cases, the corrosion fatigue is confused with heat fatigue damage.

Protection against corrosion fatigue is based on the elimination of the cyclic tensile factor, decreasing of fatigue amplitude, choice of more resistant material, decreasing of the environment aggressiveness (inhibitors) and on utilization of inorganic (metallic) coatings.

3.8 Erosion corrosion

Among the various forms of purely mechanical damage (abrasive, adhesive and erosive) in corrosive environment erosion corrosion has the greatest significance. When in contact with a flowing liquid or gas, it may lead to purely mechanical failure of metallic material and the damage grows if the flowing medium contains some species (solid or liquid particles in gas, gas bubbles or solid particles in liquid) – even under conditions when the material's mechanical failure is very small. This is

Figure 15: Erosion corrosion of stainless steel impeller.

because most of technical metals derive its corrosion resistance from passivity, i.e. the existence of a surface layer of corrosion products which restrains the anodic dissolution. In case of salt passive layers, the effect of the flowing medium may be based solely on intensive removal of the liquid layer close to the surface saturated with dissolved components of salt passive layer.

Much greater effects may be observable for turbulent flow (Fig. 15) than for laminar flow. Turbulence in a liquid may lead to mechanical failure, particularly at thicker layers, and to a substantive increase of the corrosion rate. Beyond a certain rate of flow, the corrosion rate begins to increase as well.

If the electrolyte contains particles, the attack increases even at low flow rates. Particles with sizes of the order of tens of micrometers to tens of millimeters are thicker than the passive layer, and in the area of impact they deteriorate the passive layer completely and even damage the metal below the passive layer; this gives rise to an area for new passivation of the surface. The tendency of a metal to undergo corrosion damage depends on many factors that are determined by the properties of the metal, the electrolyte, the suspension and the suspended particles as well as the geometric arrangement.

Protection against the effects of erosion corrosion is based on selection of material, on elimination of areas with unacceptable flow rates by appropriate construction solution and on removal of solid particles and gas bubbles.

With relation to erosion corrosion, the so-called fretting corrosion is indicated. It occurs under atmospheric conditions upon the surface in areas of connection of metal parts that mutually move on one another with very low amplitude. This form of corrosion occurs even under conditions of low relative humidity of the atmosphere when the corrosion effect of the atmosphere is irrelevant. Periodical removal of a thin oxide layer leads to periodical reaction of the pure surface with oxygen and following loss of material, which is transformed into corrosion products. It is usually related to textile or printing machines, bearings and other machine parts subject to vibrations. Protection is based on improving of the parts lubrication, change of material and elimination of vibrations or increasing of their amplitude.

3.9 Hydrogen damage

An important damaging process in chemical and petrochemical plants is interaction of metallic construction material with hydrogen. At temperatures below 190°C hydrogen is developed upon the surface mainly due to corrosion attended by H^+ reduction, and enters into steel in the atomic form. The hydrogen atom is notably small which enables simple diffusion into the metal structure. Generation of atomic hydrogen upon the surface of metal is also possible in water electrolytes in cathodic polarization upon the metal surface in electrochemical technologies in cathodic protection at stray current's entry.

In metals, atomic hydrogen gets caught in places identified as 'hydrogen traps', e.g. dislocations, lattice defects, grain boundaries, inclusions, blisters, cracks. Re-combination of hydrogen atoms to H_2 molecule in appropriate areas of the structure results in development of cavities and cracks. The re-combination in the areas of steel structure defects causes the development of great inner compressions (tens to hundreds of MPa), which without the presence of another stress field lead to mechanical failure (the so-called hydrogen blistering) (Fig. 16). The so-called hydrogen-induced cracking is construed as a form of blistering, when as a result of formed inner hydrogen cracks, at tensile stress of a metal part, it leads to a fracture. Loss of ductility and toughness of metallic materials is demonstrated by hydro-gen embrittlement. If the embrittled material is tensile stressed, then cracks occur (hydrogen-assisted cracking). The development of hydrides in the structure of met-als such as titanium, tantalum, niobium and zirconium greatly affects degradation of mechanical properties (hydride cracking).

Having entered the structure, hydrogen which had no opportunity for re-combination, may be removed by heating up to temperatures of 100°C–150°C. To prevent hydrogen embrittlement, it is also essential to avoid welding of moist surfaces which leads to thermal decomposition of water and entering of hydrogen into the structure.

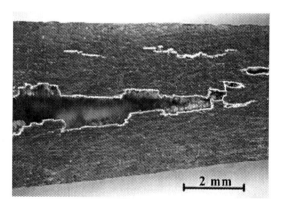

Figure 16: Hydrogen blistering of carbon steel.

The entering of hydrogen into the structure may also be possible at elevated temperatures when any aqueous environment is not present. At temperatures of $200°C$–$480°C$ molecular hydrogen undergoes thermal dissociation into atomic hydrogen on the steel surface. The atomic hydrogen diffuses into the steel, combines with carbon, and produces high-pressure methane which causes damage to the steel; this process is the so-called hydrogen corrosion (attack) which is represented by the formation of blisters.

In a proper sense, the negative effects of hydrogen on metals do not belong to corrosion processes, as an electrochemical or a chemical reaction often does not directly participate in the deterioration mechanism. The deterioration itself is of a purely mechanical type, though the causes are of a chemical nature.

4 Corrosion environments

4.1 Atmosphere

Atmospheric corrosion causes a great part of losses resulting from corrosion. The reason is that the largest surface of the construction material, mostly of carbon steel, is exposed to the effects of the atmosphere. Many metallic objects, such as building constructions, vehicles, etc are exposed to the corrosion effect of the outdoor atmosphere. Similar to the corrosion of metals in electrolytes, atmospheric corrosion is an electrochemical mechanism at normal temperatures. As the interaction of dry air with construction metals at normal temperatures is negligible, atmospheric corrosion occurs only due to atmospheric humidity. Conditions of what is known as 'critical humidity' lead to the formation of a sufficiently thick electrolyte film essential for the process of corrosion reactions. Such conditions are met where the relative humidity of the atmosphere exceeds a critical value of 60–80% (corresponds with 10–14 g H_2O/m^3 of air at 20°C). Under subcritical humidity, the corrosion rate is not zero, but for the majority of technical applications of metals it is insignificant. The corrosion effect of the atmosphere in a territorial locality is given by time for which the relative humidity of the atmosphere is above critical at temperatures of the surface electrolyte liquidity (time of wetness). Atmosphere aggressiveness is influenced by the presence of a range of substances, the most significant corrosion stimulator of which is sulfur dioxide and chlorides.

Rust, the porous precipitate of hydrated iron oxides with no especially notable protection effect, mostly visually devaluating, develops upon the surface of steel or cast iron. The protective and decorative effects of rust is exploited only in case of what are known as weathering steels, i.e. low alloyed carbon steels used in the atmosphere usually without any surface finishing, the corrosion rate of which is only a quarter to a half compared with the corrosion rate of carbon steels, and the rust layer is cohesive and uniformly colored.

The main polluting component of the atmosphere in local conditions – sulfur dioxide is oxidized by other components of the atmosphere (by atomic oxygen, NO_x) and it leads to formation of sulfate ions stimulating active dissolution of iron. Sulfates act here as a catalyzer of anodic oxidation.

In case of copper and zinc, sulfates or chlorides from the atmosphere transform into solid corrosion products (basic sulfates or chlorides), which cannot repeatedly enter into the mechanism of their active dissolution as it is in the case of iron.

Atmospheric corrosion of metals is also influenced by contents of other substances: solid substances (dust) enhance condensation (saline particles may be hygroscopic), have erosive effects; and conductive particles may work as an electrode (iron oxides, carbon) and accelerate corrosion with the effect of galvanic cells. Ammonium causes corrosion cracking of brass. Corrosion under atmospheric conditions may even be accelerated by microorganisms and accumulation of aggressive components of the atmosphere in permanently moist crevices.

Carbon steel is necessary to be protected by organic coatings or galvanizing. Modification of construction may constrain retention of moisture (orientation of profiles, elimination of heat bridges, crevices). Corrosion losses caused by atmospheric corrosion are mainly related to coating systems, their application and the necessity of their regular restoration on steel structures and objects. A frequent cause of failing of corrosion protection is application of the coating on an inappropriately prepared surface.

Protective layers formed upon copper are based on basic sulfates and chlorides (patinas). The maximum corrosion rate is approximately 1 μm per year. Wider utilization of copper in atmospheric conditions should be prevented for ecological reasons (transmission of Cu^{2+} in the living environment – the so-called run-off has been assessed to be 5 g Cu^{2+}/m^2 per year in local conditions). Copper is transmitted into the environment in concentrations of the order of ppm due to corrosion of copper roofing by draining rainwater.

Protective layers formed upon aluminum are based on modifications of Al_2O_3, or even $Al(OH)_3$. Aluminum is sensitive to pitting corrosion in the presence of chlorides. The corrosion rate is approximately 0.1–1 μm per year, and aluminum, under conditions of normal atmosphere without chlorides, is absolutely resistant. Stainless steels – if they are not exposed to chlorides – are absolutely resistant.

Exposure tests for atmospheric corrosion of construction materials are conducted at many stations in the world, and their results, together with the atmospheric pollution and meteorological data, form the basis for empirical calculation relations and for computer processing of the corrosion aggressiveness maps for given territories.

The connection of different metals leads to the formation of galvanic cells, and therefore it is necessary to avoid certain combinations, e.g. copper and iron. The washout of noble metal ions from corrosion products (e.g. Cu^{2+}) onto ions of less noble metals (e.g. Zn, Al, Fe), even if they are not electrically connected, leads to acceleration of corrosion.

4.1.1 Atmospheric corrosion of historical objects

The outer atmosphere also affects metallic monuments. Their degradation due to corrosion leads to their continuous depreciation, which requires costly restoration. The emergency conditions of copper and bronze statues are caused mainly by the corrosion of their iron framework that results in mechanical damage of a copper or a bronze shell. Construction defects of statues allow easy leakage of rainwater to the inside, and conversely, create conditions for internal retention of water. Good electrical and electrolytic contact of the framework and the copper shell aids the corrosion damage of the iron parts as well.

With regard to corrosion resistance of iron, ill-informed literature and media often cite the example of the so-called 'non-corrodible' Iron Pillar in Delhi, though iron corrosion products apparently cover its surface. This column, made of malleable iron (length 7.2 m, weight approx. 6 tons), is an evidence of the admirably high technical skills of the inhabitants of India 1600 years ago. In its aboveground part, the column corrodes very slowly, mainly owing to the clean atmospheric conditions that existed in Delhi in the past and which are still unfavorable for the long-term condensation of aerial humidity. The great heat capacity of the column, given by its weight, ensures that corrosion occurs very seldom, even in times when the relative humidity of the atmosphere exceeds the critical value. On average, the percentage of rainy days in a year is only approximately 10%. The thin layer of corrosion products formed during the periods when the atmospheric aggressiveness was lower compared to the rainy days has protected the column till date. The underground part of the column has obviously undergone corrosion. The composition of the column is well known, and considerable variation has been observed based on the area of the sampled specimen owing to the production technology (malleable iron) used. To a certain extent, the composition contributes to the atmospheric resistance of the column as well (low content of sulfur, higher content of phosphor). The rust structure partly resembles the rust developed on weathering steel. According to some sources, the higher content of phosphor (tenths of wt.%) was achieved on purpose, as during production a certain wood with higher content of phosphor was used. Specimens of the metal sampled from the column did not exhibit any significantly better corrosion protection in comparison with other carbon steels. With respect to the current scientific knowledge, neither corrosion nor technological mystery is related to the column in Delhi.

4.2 Aqueous solutions

4.2.1 Water

The term, water, for our purpose includes not only chemically pure water, but also various weak-concentration aqueous solutions with a content of substances that get into water during its natural and industrial hydrological circle. Not only do natural waters aggressively affect water constructions and ships; their aggressiveness passes into the industrial waters obtained from them and also into drinking water.

The aggressiveness of industrial waters is influenced either by intentional modifications or by the substances that get into the water during its use or that concentrate during its circulation. Although there are several requirements imposed on various kinds of water, there is a range of common factors influencing the aggressiveness of the water.

The metal surface exposed to absolutely pure water corrodes only irrelevantly. Mainly additional agents, a certain amount of which are always present in both natural and industrial waters, are responsible for the process of corrosion reactions in the water.

Water aggressiveness is affected by the content of oxygen and other aggressive gases, by the amount and type of dissolved salts, by the presence of organic substances and microorganisms, by pH, temperature and flow rate, and by the content of solid particles.

Oxygen is the most influential of gases dissolved in water as corrosion in water is mostly controlled by the rate of oxygen reduction. The uniformity of the oxygen access to the surface is also very important – if the access is not uniform it leads to the formation of cells with differential aeration that results in local attack.

Oxygen is not only a stimulator of corrosion, but also supports formation of protective layers and if its content in water is sufficient, it may limit the corrosion. However, the content of oxygen dissolved in pure water that is in contact with a free atmosphere at room temperature (up to 10 mg O_2/l) is not sufficient for passivation of carbon steel.

Carbon dioxide is another important component influencing aggressiveness of water. Thus, in acidic waters, cathodic reactions of types other than oxygen reduction may apply. The presence of CO_2 also negatively influences the inhibitive effect of Ca^{2+} and HCO_3^- ions. Similar effect is also caused by sulfur dioxide dissolved in water from the polluted atmosphere. Hydrogen sulfate and ammonium may cause corrosion cracking.

Salts dissolved in water change the pH of the solution by hydrolysis, increase conductivity, influence formation of layers upon the metal surface and may be a cause of localized corrosion.

Chlorides are present in almost all types of water and cause problems mainly when stainless steels are used, such as pitting and crevice corrosion, and also corrosion cracking at higher temperatures. Aluminum is also easily attacked by pitting corrosion due to the effect of chlorides in water. With the exception of seawater and brackish water, the usual concentration of chlorides in water lies within the range 10–200 mg Cl^-/l. Compared with chlorides, the corrosion effect of sulfates and nitrates on steels in normal concentration is low.

Natural and industrial waters have pH usually ranging from 3 to 9, and therefore on a majority of metals, a layer of insoluble corrosion products is formed. The quality of this layer often determines corrosion resistance.

In the formation of a protective layer, the hydrogencarbonate, calcium (II) and magnesium (II), ions, with their inhibitive effect for steel given by the ability of aerated water to form protective layers consisting of iron oxides and calcium carbonate, constitute a significant component of water.

In power plants, water is used for cooling and heat transmission in armatures of district and central heating, as hot water in open circuits and is supplied to boilers. Water for the production of steam, upon which special requirements are imposed, is treated to the lowest salinity by demineralization, depleted of oxygen (thermally, by hydrazine) and alkalized with sodium phosphate. With water supply depleted of oxygen, corrosion of a boiler steel surface depends at higher temperatures on the formation of well-protecting layers of oxides (Fe_3O_4). The protective layer of magnetite is damaged by higher pH and oscillation of temperature.

Power plant cycles coming into contact with supplied water are most frequently made of carbon steel, brass and stainless steel. Brass, with a higher content of Zn, is affected by dezincification. At elevated temperatures, in the presence of chlorides, stainless steels are affected by corrosion cracking.

It is necessary to soften and deaerate water for heat transmission in closed circuits. Water for armatures of central heating is modified by addition of inhibitors. For the water in closed power plant cycles, it is important, much as in the case of water supply, not to let the water get into contact with the atmosphere, not to change the water and not to replenish it, as corrosion depends mainly on the amount of oxygen entering the loop. In this case, oxygen is depleted by addition of sulfite. For the distribution of utilized warm service water, drinking water is sometimes modified by adding silicates or phosphates. If the water is sufficiently conductive, it is possible to cathodically protect certain steel parts of the system (boilers, tube sheets).

Cooling water used in the circuit systems of industrial plants (e.g. chemical) and in power plant cycles are almost always aerated, as it is in contact with the atmosphere in the cooling towers. Therefore, corrosion may not be limited by the decrease of the content of oxygen. Where necessary, the inhibitors (phosphates) and other substances, which constrain the activity of microorganisms (biocides), keep the insoluble substances floating and limit deposition of calcium carbonate even if the solubility product is exceeded, are charged.

Cooling water for flow through systems is cleansed of dirt and is not treated any further. To constrain formation of deposits, the appropriate flow rate for water distribution systems is from 1.5 to 3 m/s. For erosion corrosion, the appropriate maximum flow rate is 1.5 m/s.

The equipment coming into contact with cooling water is made of carbon steel, stainless steels or brass. Carbon steel may be protected by coatings. Brass may suffer from dezincification and, in the presence of ammonia, also from corrosion cracking. In the presence of greater amount of chlorides in water (>100 mg Cl^-/l), stainless steels may crack and is affected by crevice and pitting corrosion.

In the mine water environment, increased aggressiveness is caused not only by the presence of anions (chlorides, sulfates, carbonates), but also by cations which may, for instance, participate in cathodic reaction (copper deposition).

The aggressiveness of drinking water depends on the composition of natural water out of which it is modified. Drinking water should have an equilibrium composition and normally have a content of salts 300 ± 200 mg/l and chlorides 10 to 100 mg/l. Equipment made of carbon steel, galvanized carbon steel, brass and copper are used for drinking water. Carbon steel is often protected by organic coatings (paints). In the drinking water-distributing armatures, extensive iron corrosion products frequently occur under which it leads to localized corrosion.

It is essential to note that in warm waters (service, cooling), the zinc coating protects carbon steel only up to 50°C; in warmer waters carbon steel becomes nobler than steel and results in the formation of macrocells of the small anode (steel in the area of the coating defect) and the large cathode (zinc coating).

The combination of copper and galvanized steel or aluminum in the same circuit is also inappropriate. When copper, which is transferred in very small amounts to the solution due to corrosion, comes into contact with a galvanized steel or aluminum surface, it deposits resulting in the formation of local cathodes which cause local attack.

The aggressiveness of natural waters is very variable and depends not only on contents of substances naturally present in water, but also on the grade of pollution caused by the humans. The corrosion resistance of the same material therefore differs in different localities, and the aggressiveness of water is determined by chemical analysis.

Seawater contains on average 35 g/l of dissolved salts (mainly chlorides) and is alkaline (pH 8) and well conductive. In the coastal areas, seawater is used for industrial cooling, but the main corrosion problems occur in ships and water constructions. In seawater, carbon steel, which is protected by coatings or cathodic protection, is used again as the main construction material. As to metallic materials, titanium, nickel and its alloys exhibit a very good resistance. Stainless steels suffer from pitting and crevice corrosion and stress corrosion cracking. Corrosion problems in brackish waters (fresh water polluted with seawater) are similar to that observed in seawater.

4.2.2 Industrial electrolytes

The aggressiveness of industrial electrolytes depends on the kind of anions and cations present in the solution, and it is impossible to completely identify it by rules generally valid for different chemical substances.

Acids are characterized by a low value of pH, and therefore, evolution of hydrogen is the main cathodic reaction in the corrosion of most of technically utilized metals. If oxygen is present, its reduction also plays a role. The aggressiveness of acids largely depends on the type of anions and the presence of substances which may have oxidation effects or may attack the passive layers (Cl^- in stainless steels, F^- in titanium).

The presence of oxidizing agents is harmful for non-passivable metals under given conditions. For passivable metals, such as stainless steels, the presence of oxidizing agents is favorable, but only when their concentration is adequate. Positive influence is also reported for anions that show no oxidative effect. Their positive influence is thus based on formation of protective insoluble products.

After the dissociation of salts in solution, both cations and anions may participate in the electrochemical processes related to corrosion of metal. In case hydrolysis of salt occurs, the pH of the solution is also influenced. If the dissolution of salt results in acidic solution, its aggressiveness corresponds to the solution of acid with the relevant anion of the same pH. The case is similar with alkaline solutions, the aggressiveness of which corresponds to solutions of relevant hydroxides. The effect of individual ions may be inhibitive, stimulating or indistinctive. Both ions (cation and anion) may act negatively as well as positively, or against one another. Similar to other environments, the basic as well as accompanying solution components with oxidation effect are of great importance.

Concentration of salt solutions has an obvious effect on their aggressiveness. Along with increase of concentration (up to a certain limit), corrosion aggressivity also increases. Sometimes it is related to increase of solution conductivity and intensification of the cells function. The corrosion rate in neutral salt solutions is mostly controlled by the rate of oxygen reduction. Corrosion of non-noble metals proceeds in alkaline environment with hydrogen evolution. For noble metals, oxygen reduction is decisive.

Occurrence of certain forms of localized corrosion (pitting, corrosion cracking) is related to the presence of certain ions. Of all anions, chlorides cause probably the most harmful corrosion problems as they accelerate the uniform corrosion and also cause development of crevice and pitting corrosion and corrosion cracking.

4.3 Soil

Most of the metallic structures (tanks, pipelines) buried in the soil are made of steel. Against the effects of soil electrolytes it is protected mainly by coatings based on tar and bitumen, or more perfectly by polymeric coatings (polyethylene, polypropylene, etc.). This protection is usually completed by cathodic polarization. The selection of optimal protection depends on the significance of the equipment, aggressiveness of soil and the type of protected metal. The type and cohesion of the soil, the homogeneity, humidity, chemical composition of soil electrolyte (including gases), pH and redox potential, buffer capacity and oscillation of ground water level are significant for aggressiveness of the soil. Corrosion attack of metallic materials in a soil would be absolutely indistinctive without the presence of humidity.

In addition to the depolarizing effect (cathodic reduction), oxygen supports decomposition of organic substances, oxidation of sulfides up to sulfuric acid. Carbon dioxide negatively affects spontaneously generated protection layers (so-called calcareous rust consisting of hydrated iron oxides and calcium carbonate). Aqueous solutions of carbon dioxide are claimed to attack ferritic–pearlitic steels along the boundaries of the grains. Metal corrosion induced by aqueous solution of carbon dioxide at pH 4.5–5 is greater than the corrosion induced by aqueous solution of HCl at equal pH. Aqueous solutions of CO_2 also stimulate selective corrosion of gray cast iron – graphitic corrosion. In the deeper layers of the soil, the concentration of CO_2 is considerably higher than it would obtain in the composition of air.

In most of the soils, pH ranges from 5 to 9. The dividing value is 6.5, below which aggressiveness is high; above 8.5, the soils are deemed as unaggressive. However, under mechanical stress, increased alkalinity at higher temperatures (exceeding 50°C) may commence attack along the grain boundaries. Besides, increased alkalinity may result in serious defects of amphoteric metals (aluminum, lead).

From the corrosion perspective, the presence of soluble salts is always unfavorable as they increase conductivity and thus enable function of macrocells, retain humidity and have activating effects. Cl^- (road salting in winter, soil in coastal territories), Na^+, Ca^{2+} and Mg^{2+} occur in soils. Ions of Cl^- in the presence of Mg^{2+} shift pH to lower values (hydrolysis). A similar but less significant effect is also exhibited by SO_4^{2-}. The presence of S^{2-}, HS^- is unfavorable at any time, as it considerably stimulates the corrosion effects.

To assess the aggressiveness of soil, the value of apparent resistivity is used: this is sometimes the basic criterion of aggressiveness. Resistivity is determined in corrosion analysis in natural conditions and constitutes a basis for application of cathodic protection (cathodic protection is necessary at values below $10\,\Omega m$). The soils are considered as unaggressive at resistivity exceeding $100\,\Omega m$.

Corrosion defects of buried structures occur in areas of damaged insulation, more so with insufficient cathodic protection. Soil aggressiveness is further increased by microbial activity and presence of stray currents, and alternatively by excessive polarization resulting from incorrectly installed cathodic protection and by increased temperature.

Metallic constructions in soil conditions are frequently exposed to stray currents which may lead to their quicker failure if they are not sufficiently insulated, are buried incorrectly in relation to the source of stray currents, and not protected against stray currents by means of electric or electrochemical protection.

Coating is the most widespread means of corrosion protection against soil affecting steel pipelines and other buried structures. The efficiency of this protection is mainly influenced by adhesion to metal, slight decrease of the coating resistivity even after long exposure to soil or water, low coating's inclination to swelling and ageing, impermeability for water and gases even at higher temperatures, and resistance against mechanical stress. No coating completely meets the abovementioned requirements, but when polymers were introduced in coating techniques as a substitute of traditional bitumen insulations, resistance significantly improved. Cathodic protection which secures protection in areas of damaged insulation serves as a complementary protection.

4.4 Concrete

The required mechanical properties are usually reached with use of reinforcing bars which in most cases are made of carbon steel. Carbon steel is passive in fresh concrete as after the intrusion of water, the reserves of calcium hydroxide produce pore solution with pH of 12.5–13.5 (depending on the contents of alkali metals in cement). Such alkalinity ensures spontaneous passivation of steel and very low corrosion rates. The oxidizing power of the pore solution is given by oxygen accessibility from the atmosphere. The pore solution corrosion aggressiveness depends on the access of the carbon dioxide and chlorides to the reinforcement. When carbon dioxide from the atmosphere intrudes into moist concrete, free calcium hydroxide reacts and turns into calcium carbonate (so-called concrete carbonation), which is accompanied by decreasing pH of the pore solution.

In case the carbonation front gets to the reinforcement through the covering layer (20–30 mm), it causes activation of the steel and substantial increase in the corrosion rate. The rate of the carbonation front progress depends on quality of the cement and concrete. The progress rate (in order of tenths of mm per year) depends not only on the density of concrete and on the presence of cracks, but also on concrete moisture. The pH decrease may be accelerated by the effect of acid rain.

In case the chlorides penetrate into concrete (even if non-carbonated) it also causes activation of the steel without necessarily being related to decrease of the pore solution pH. The critical concentration of Cl^- in concrete is approximately 0.2 wt.%/cement. The chloride penetration through concrete is in the order of mm per year.

The steel reinforcement which is exposed to the effects of the outer atmosphere prior to being installed into concrete has a surface covered with rust and its corrosion rate, when it is embedded into the concrete, achieves its bordering limits as concrete alkalinity does not entirely inhibit the rate of active dissolution under the layer of rust. Therefore, it is inappropriate to install pre-rusted reinforcement into concrete; scaling of reinforcement from production has actually no major effect on carbon steel corrosion resistance.

Where steel reinforcement is activated by the effect of carbonation or penetration of chlorides, its corrosion rate is given mainly by transport of oxygen through the concrete covering layer or cracks in the concrete.

Should the concrete be in a passive state, the corrosion rate is lower than $0.1\,\mu m$ per year and the corrosion is uniform. In active state, the corrosion rate exceeds $10\,\mu m$ per year – the activation is localized. Corrosion damages are usually non-uniform due to cells with differential aeration.

From the technical point of view, the corrosion rate of $1–2\,\mu m$ per year is acceptable as it corresponds to the 100-year service life of a reinforced concrete structure.

Corrosion damages result in occurrence of voluminous corrosion products (2–3 times greater than the volume of the parent metal), which will cause cracks in concrete and, concurrently, reduction of the reinforcement cross-section (acceptable reduction shall not exceed 10%).

To extend the period before the steel reinforcement activation, it is necessary to utilize the highest-quality concrete (water-to-cement ratio <0.5, sufficient contents of quality cement, chloride-free components for concrete preparation).

The basic methods of concrete reinforcement corrosion protection are:

- Limited intrusion of moisture into the concrete (insulation, concrete surface coating).
- Limited intrusion of corrosion stimulators into the concrete (Cl^-, H^+, O_2).
- Reinforcement coatings (organic, zinc).
- Corrosion inhibitors.
- Alternative materials used for reinforcement (e.g. stainless steels)
- Electrochemical methods (cathodic protection, re-alkalization, chloride extraction)

Powder epoxides are the most reliable and commonly used organic coatings in the world. The price of the reinforcement is about to increase 1.5–2 times. However, organic coatings only extend the service life of a reinforced concrete structure and do not ensure a long-term service life. Problems are caused by additional bending and connecting of the reinforcement coated with the organic coating. All defects occurring during these operations and other improper treatment must be repaired so that maximum reliability is ensured.

Hot-dip galvanized coatings have thickness of 35–100 μm (proportionally to the thickness, the corrosion resistance increases). The zinc layer eliminates the effects of carbonated concrete; however, the corrosion rate in fresh concrete is still very high. The corrosion reaction is accompanied by development of gaseous hydrogen, which negatively affects the structure, and thereby also the mechanical properties of the concrete interfacing with the reinforcement. Development of hydrogen is not effectively eliminated even by practically utilizable conversion coatings of zinc. Zinc coating doubles the price of reinforcement.

Stainless steels are entirely resistant in carbonated concrete, but concrete carbonation leads to reduction of critical chloride concentration. In the presence of chlorides, there is a danger of pitting or crevice attack. Austenitic chromium-nickel steels (FeCr18Ni10) are suitable in most of the practical cases of chloride contamination. Austenitic chromium-nickel steels containing molybdenum (FeCr17Ni12Mo2) are suitable (including cleaned welding seams) under all conditions. Duplex steels (FeCr23Ni4 and FeCr22Ni5Mo3N) are suitable not only for corrosion resistance, but also when high mechanical strength is required. A decisive factor of corrosion resistance of stainless steels is removing the scales occurring after production or welding of the bars. Chromium stainless steels (FeCr$_{13}$, FeCr$_{17}$) are resistant in chloride-free carbonated concrete. However, in concrete contaminated with chlorides, their resistance after removal of scales is higher than in case of scaled chromium-nickel steels. The price of the reinforcement made of stainless steel is about 5–10 times higher compared to carbon steel: the increase in the price of the structure shall not exceed 10%.

Long-term protective effect of corrosion inhibitors is questionable, particularly from the practical point of view.

The corrosion resistance of the steel reinforcement in moist concrete may be significantly affected by direct current passage during which a change of the electrolyte composition takes place that is favorable for spontaneous passivation of steel (increase of OH^- concentration and decrease of Cl^- concentration). The electrochemical methods of concrete reinforcement protection have been used for many years. In particular it concerns the cathodic corrosion protection where the protected reinforcement is permanently connected to the negative pole of a direct current source or to a sacrificial anode. Apart from this method, procedures of non-recurring cathodic polarization are also used, including electrochemical realkalization and electrochemical extraction of chlorides.

Cathodic protection in concrete is usually divided into two groups based on whether it is installed preventively or as an additional corrosion precaution after activation of the reinforcement. The preventive cathodic protection in a new unaggressive concrete requires very low protective currents and impedes chloride intrusion (migration acts against Cl^- diffusion) and even carbonation (OH^- occurs due to oxygen reduction on the reinforcement). Cathodic protection in an aggressive concrete requires increased protective currents and contains both chloride extraction and re-alkalization.

A very important supporting measure against corrosion of steel reinforcement in concrete is corrosion monitoring.

5 Corrosion protection methods

There are four basic groups of corrosion measures which may decrease the corrosion deterioration of metal:

- material selection (metallic, inorganic non-metallic, polymeric),
- coatings and surface treatment,
- modification of corrosive environment (change of physical parameters, destimulation, and inhibition),
- electrochemical protection (cathodic or anodic protection), which is followed by the fifth group of corrosion measure
- corrosion prevention by design.

The selection of the means of protection, or more often of their combination, usually depends on many circumstances. Nevertheless, at any time it is essential to select construction material and design, and often also an appropriate surface treatment. Electrochemical protection and modification of environment belong to complementary or specialized corrosion procedures.

5.1 Material selection

Material selection is a very significant means of corrosion protection, the objective of which usually is to replace carbon steel that in most cases would meet the requirements of primary utilization qualities. The most distinctive material developed for corrosion protection are stainless steels. It is a group of iron-based alloys containing chromium (>12 wt.%) and a range of other elements (Ni, Mo, Ti, Cu, N). From the corrosion perspective, there are three main groups of stainless steels – chromium (martensitic and ferritic), chromium-nickel austenitic and duplex. The stainless steel most exploited (70%) is austenitic steel (FeCr18Ni10). In case stainless steels are high-alloyed (Cr, Mo), they are referred to as superaustenitic, superduplex or superferritic. The nickel alloys with chromium exhibit even higher corrosion resistance. The greatest corrosion resistance in many environments is achieved by the use of titanium and its alloys or tantalum.

The use of copper and its alloys, lead, zinc and aluminum may not be omitted in corrosion protection; however, it is not that universal, as it is related to specific conditions. For substitution of metals, use of polymers, glass and ceramics, or graphite is very significant.

5.2 Coatings

The most widespread procedure of corrosion protection is application of coating onto the basic metallic material which is not sufficiently resistant in a given environment. Carbon steels are the most coated metallic material. Ratios among various surface treatments are the following (according to the area of protected surface): 70% organic coatings, 20% metallic coatings, 10% other types (conversion and cement coatings, enamels).

The basic protection mechanism of coatings is to create a barrier between protected metal and corrosive environment. Depending on the type of coating used, protection side effects may subsequently occur, e.g. galvanic, destimulating and inhibitive.

Among materials of organic types used in corrosion protection are rubber and other polymer coatings, in addition to most frequently used coatings,. Materials for production of organic coatings are mixture of polymer substances and special additives (inhibitors, pigments, fillings, etc.). Their protective function is in the nature of a barrier. By selecting appropriate additives it is possible to exploit their inhibitive or destimulating effects (Zn). Organic coatings are used most frequently for corrosion protection of metals. The efficiency of this form of corrosion protection is related not only to the paint quality, but also to the quality of prior surface treatment and the way of its application. Painting systems have a limited service life; therefore, they must be renovated, and they thus constitute the greatest part of total costs incurred by corrosion protection.

The protective effect of metallic coatings depends on the nature of formed composite, that is, on the mutual corrosion resistance of a coating and the basis in a certain corrosive environment. Various protective mechanisms are applied on imperfect coatings. Coatings produced without defects exhibit a basic effect that is in the nature of a barrier. The existence of defects in metallic coatings and the exposure of such area in corrosive environment allows both the materials to perform different electrochemical behavior. If the coated metal is of a more positive free corrosion potential than the coating, it leads to corrosion of the coating and protection of the basic metal (the coating serves as a sacrificial anode). If the corrosion products of the coating metal are insoluble, they continuously fill in the defect of the coating. The presence of pores in this type of coatings is (to a certain extent) not decisive. In case the coated metal is less noble than the coating, its protective function is merely in the nature of a barrier. In case of the failure of coating compactness, cells are formed in which the exposed basic metal forms a small anode (i.e. by corroding area), and the surface of coating metal forms a large cathode. Corrosion of the coating metal often results in undercorrosion of the coating. It is very difficult to achieve zero-defect coatings; therefore, coatings with several layers of coating metals are used.

In a chemical treatment the metal reacts with the treatment agent so that a coating not easily soluble is formed on the metal surface. Examples of chemical conversion treatment are phosphating, chromating and anodizing.

Coatings of a cement nature constitute a specific group of inorganic coatings. They are mostly used for protection of steel surfaces; the most remarkable advantage of their use is the possibility of application directly onto untreated (corroded) surface. The protective effect is in the nature of a barrier as well as destimulating. Enamels belong among significant anticorrosion coatings.

5.3 Modification of corrosive environment

Corrosion processes may be reduced by modification of corrosive environment. The objective of this procedure is to change the concentration of the component causing the attack, or to remove it from the environment – destimulation (removal of oxygen, chloride ions, air humidity, etc.), modification of physical parameters (temperature, flow rate), or inhibition of corrosion reactions (i.e. addition of corrosion inhibitors).

Destimulation is usable in cases when the aggressive component of the environment is not necessary for the production technology, and its elimination does not require excessively high expenses. Destimulation thus is mostly used in cases when the corrosive substance is present in a relatively low amount. The most common case of destimulation is removal of oxygen from an environment in which the corrosion process is regulated by the rate of oxygen reduction. The decrease of oxygen content of water that is used in power plants may be achieved by either physical methods (vacuum, boiling, bubbling with inert gas) or chemical processes (reaction with hydrazine or sulfite). The water treatment is also possible using an ionex or electrochemically. The alkalization of acidic waters or removal of chlorides from water by demineralization on ionex is also considered as destimulation. The chlorides are removed mainly in cases when they might cause pitting and crevice corrosion or corrosion cracking of stainless steels.

The presence of aqueous electrolyte is a basic presumption of a course of the corrosion process; elimination of humidity, e.g. from organic products of solid salts, or from the atmosphere in closed rooms, leads to significant reduction of corrosion. Removal of solid particles from aqueous electrolytes and air constrains attacks related to formation of deposits (crevice corrosion in solutions and the atmosphere) and to erosion-corrosion effects of the environment.

In some cases, corrosion may be limited by small changes of concentration of the substance which causes the corrosion attack. It may be influenced also by a change of ratio of concentrations of substances with inhibitive and stimulating effect. This is mostly the case when conditions for spontaneous passivation of metal or stability of a passive layer are secured by this modification.

Particularly, changes of temperature and the environment flow rate belong to the group of physical parameters modification. Lowering of corrosion aggressiveness may be achieved by decrease of temperature. In some cases, corrosion is conversely suppressed by increasing the temperature which, for instance, leads to decrease of oxygen solubility and the steel corrosion rate in water. Increase of temperature may have a positive effect in reducing atmospheric corrosion in closed rooms, as the relative humidity may drop below the critical level.

Corrosion attack may also be influenced by adding of substances (corrosion inhibitors) into the corrosive environment, which causes its considerable decrease in low concentrations. The effective concentration of these substances usually ranges from 0.1 to 10 g/l. In industrial conditions, the inhibitors are determined mainly for acid pickling of metals (e.g. thiourea, urotropine), recirculating cooling water systems in industrial plants (e.g. polyphosphates, silicates, phosphonic acid) and protection of equipment used for oil production and refining (e.g. imidazoline, amines, thioglycolics acids). A range of inhibitors is determined also for protection of machinery products against atmospheric corrosion.

The volatile corrosion inhibitors, i.e. substances with inhibitive effect and high tension of vapors (amines, organic nitrites) by which the air is treated or packing materials are impregnated, also protect against atmospheric corrosion. The inhibitors soluble in lubrication oils (an inhibitor may be contained in oil emulsions for machining) are determined for protection of machinery products as well. Even organic coatings determined for temporary protection frequently contain substances with inhibitive effect.

5.4 Design

The design of an object, which is exposed to a corrosive environment, should take into consideration the requirements of its effective corrosion protection. It is thus often needed to modify its design, i.e. change its dimensions or shape, or use specific production procedures during mechanical operations. Design changes may affect the corrosion resistance of equipment so that they may cause changes in the properties of the material, influence the quality or effect of corrosion protection and have an indirect effect on the aggressiveness of the environment and on time during which the surface is in contact with the electrolyte.

Machinery operations, such as welding, soldering, mechanical working and machining may significantly decrease resistance against corrosion even of the construction material that might in itself be sufficiently corrosion-resistant. When welding, a change in the structure of the basic metal that causes a disposition to the localized corrosion may occur. Corrosion problems ensuing from the quality of weld seams may be resolved by selection of convenient combination of basic and filler alloys, welding technology, and design of the weld seam. Soldering, similar to welding, may cause changes in structure, formation of cells and crevices.

While producing equipment with weld seams, cold working or working on the machined surfaces, internal strains occur which may cause environmentally induced cracking. Machining also affects the roughness of the surface, and since the corrosion resistance of smooth surfaces is mostly higher, it is appropriate to use technologies that produce surfaces of the lowest roughness.

Overall corrosion also depends on the area of the exposed surface; therefore it is necessary to design equipment with the smallest possible area of exposure (one big vessel is better than more small vessels). Increase of a metal material thickness, the so-called corrosion allowances that, under given uniform corrosion and with correct presumption of its rate, ensure a sufficient cross-section of the material during the estimated service life of a object, may be deemed as a constructional measure as well. The design must also allow to change the parts attacked more quickly by corrosion.

Coatings generally require the protected surface to be as simple as possible and to be of such shape and dimensions as allow application of a uniform layer all over the surface. Seams that are to be coated are preferred to be welded or soldered. It is important that such design is selected so that it does not lead to unnecessary mechanical damages of the coating during the operation. Certain kinds of coatings (e.g. enamels) require a minimum thickness of the basic material. Design must ensure maintenance of those coatings, the service life of which is shorter that the service life of the equipment. Mainly paints are concerned here. The whole exposed surface is to be sufficiently accessible both for preparation and re-application of the coating.

Reduction of the time during which the surface is exposed to corrosive environment is based on use of devices that do not retain the corrosive environment, or enable a complete emptying. In the outdoor atmosphere, it is vital to use simple structures in which the profiles and their connections are so arranged that rainwater is not retained.

Inhomogeneities on the exposed surface may stem from connections of two differently noble metals – for example, during welding and when the basic metals of different compositions are connected, a macrocell occurs. Similarly, inhomogeneities can occur when the galvanized connecting material is combined with the non-galvanized one, or when two adjacent parts of the equipment are of different metals.

Should a coating be used for elimination of a macrocell, it is to be used on the cathodic material, in particular, since the damaged areas act as small anodes with concentrated corrosion attack when the coating is damaged – which unavoidably happens from time to time. From this perspective, it is more appropriate to apply the coating only on more resistant materials.

Another factor affecting corrosion resistance is the flow character; its size and uniformity are often determined by the object design. Problems appear, for instance, in the area of pipe bends. Areas where the flow rate exceeds acceptable limits for a given material, and where sudden changes of flow occur should be avoided.

5.5 Electrochemical protection

Electrochemical corrosion protection is based on the effects of direct current passing through a protected metallic surface which lead to decrease of corrosion rate. Passage of current results in occurrence of a change of metal electrode potential as well as a change of composition of corrosive environment adjacent to the protected surface. If the changes in the environment composition are insignificant, then it is, in case of cathodic protection (the protected object is a cathode), a reduction of anodic dissolution of metal in activity, and in case of anodic protection (the protected object is an anode), a current-induced passivation. In case of cathodic protection it leads, under conditions of limited convection, to alkalization of the environment at the protected surface and creation of conditions for the stability of passive state (mainly of steel which is the most cathodically protected metal), even under conditions of cathodic polarization. The stability of passive state is also ensured by depletion of chloride ions by migration off the cathodically protected surface. The cathodic protection in the area of stabile passivity is also used for elimination of great oxidation ability of the environment which in case of stainless steels leads to pitting corrosion. Cathodic polarization may be used for elimination of aggressiveness of porous materials, such as carbonated or chloride-contaminated concrete (re-alkalization, extraction of chlorides). The effects of direct current on corrosion resistance of metals may also be negative. It may be observed in the case of improperly operated electrochemical protection and in case of the so-called stray currents, which lead to undesirable polarization of the surface that is in contact with the electrolytically conductive environment.

5.5.1 Cathodic protection

Cathodic protection is mostly a complementary means of surface protection of steel covered with coating. Under certain conditions, such as in seawater, cathodic protection may be used also on uncoated metallic surfaces.

Cathodic polarization of protected metallic surface may be generated by two means: connection of the metal (sacrificial anode) with more negative free corrosion potential in a given environment than the required protection potential (in this case the effect of galvanic cell is used) or connection of the protected metal with the negative pole of direct current supply.

Carbon steel is the most frequently cathodically protected metal in aqueous environment (soil electrolyte, industrial and natural waters). Under these conditions, magnesium, zinc and aluminum may be used for protection. The criterion for use is that the free corrosion rate of the anode is lower so that it is not passivated (therefore aluminum is used mainly in seawater). Localized effect and possibility of usage without any electrical sources are advantages of a sacrificial anode. Replacing the sacrificial anode is disadvantageous. Sacrificial anodes are exploited for protection of ship hulls, boilers for water heating, steel structures exposed to seawater and locally, also of buried structures.

Breaking of greater ohmic resistance and protection of large objects with a small amount of anodes are enabled by cathodic protection using an external source of direct current. However, in this case it is necessary that the operation of protection does not lead to degradation of the anode by corrosion. Therefore, usually those materials are exploited that have electron conductivity and are anodic dissolution-resistant – the main anodic reaction is evolution of oxygen or chlorine (in chloride environment). Silicon cast iron, graphite, magnetite, titanium with platinum or iron oxides coating are suitable anode materials.

For cathodic protection by external source of current it is important to control so that polarization of the surface to the area, where hydrogen is liberated by decomposition of water and the environment is intensively alkalinized, does not take place. It may negatively affect further damages of protective coating of the protected structure (disbonding), as well as steel properties with the entering of atomic hydrogen.

5.5.2 Anodic protection

Anodic protection is based on intentional passivation of metal by current passage or change of a metal potential within the scope of passivity by current passage when a protected object is an anode. Anodic protection constitutes a primary corrosion measure taken in strongly aggressive industrial solutions, e.g. sulfuric acid.

5.5.3 Protection against stray currents

If direct current passes through an electrolytically conductive environment, it results in drop of potential in the metallic structure that is in contact with the environment which affects the corrosion potential. It leads to formation of anodic and cathodic areas. The corrosion rate increases mainly in anodic areas where a wall may be perforated. Stray currents most frequently affect structures placed in soil, water or in industrial electrolytes. Stray currents in soil may be caused by direct current transportation systems (railways, trams, underground), stations of cathodic protection (interference), electrochemical technologies (galvanizing plants, plants producing sodium hydroxide and chlorine), telluric currents (currents in earth crust generated by the effect of induction from magnetic resonance between ionosphere and the earth) and ore body (currents caused by activities of macrocells on ore body).

In line-buried structures (pipelines, cables), stray currents may be limited by several means: measures taken against the source of stray currents (decrease of resistance for back current in direct transportation systems), insulating a metal subject from soil and installation of complementary anodes for the systems of cathodic protection. The limitation may also be achieved by suitable selection of the line equipment route, improvement of line equipment insulation, use of insulation flanges and by drainages, electric connection of the stray current source and the threatened structure. Controlled rectifiers for cathodic protection may also be exploited.

5.6 Corrosion testing and monitoring

Although we are coming to know more and more about the patterns regulating the corrosion processes and are able, to a certain extent, to anticipate behavior of materials in the environment, it is still essential to experimentally obtain most of information on corrosion resistance.

The reason is the complexity of the corrosion process and the possibility of this process to be affected by a range of random factors. Under laboratory or operational conditions, experimental works for obtaining information on corrosion may be summarized in terms of corrosion tests. Some kind of a corrosion test is always related to the purposes of its execution, the type of corrosion system (material/product/environment) and the form of corrosion which causes possible or factual deterioration. A range of laboratory and industrial corrosion tests procedures are standardized.

The principles of corrosion tests are based on evaluation of visual, dimensional and mass changes; metallographic evaluation, evaluation of changes in mechanical properties, evaluation of changes in the corrosive environment composition and analysis of corrosion products are exploited. A significant part of it are electrochemical methods that are based mainly on dependence between corrosion potential and the current passing through the metallic surface and their time dependence. In corrosion tests, other physical methods are also exploited, such as resistance, inductive, acoustic, radiation, magnetic, thermographic procedures, etc.

Corrosion monitoring is important to determine operational reliability of equipment. It is based on technologies that help to obtain relatively fast and continuous, if possible, information on corrosion. A brisk response on acceleration of corrosion is needed to be able to intervene immediately. Corrosion monitoring concerns not only industrial equipment but also reinforced concrete structures, and is also used for tracking atmosphere aggressiveness.

6 Economic impacts

In industrial countries, direct corrosion costs reach approximately 3–5% of the gross national product. The latest evaluation of direct corrosion costs in the USA annually makes for US$ 276 billion. This immense loss, which is 50 times higher than direct annual losses caused by fires, is attributed mainly to the fact that corrosion is not completely avoidable, but also to the fact that the existing technical potentialities of corrosion protection are insufficiently exploited. The types of corrosion costs vary from direct losses on material to hardly assessable environmental and emergency impacts.

It is believed that at least one-fourth to one-third (in some fields up to 70%) of losses caused by corrosion may be avoided. Indirect costs due to metallic corrosion may be substantially higher than direct losses and may have considerable health and environmental impacts (e.g. damages of buried equipment, damages in nuclear power stations or chemical plants).

Literature for further reading

[1]   Pourbaix, M., *Atlas of Electrochemical Equilibria in Aqueous Solutions*, NACE: Houston, TX, 1974.
[2]   Landrum, J.R., *Designing for Corrosion*, NACE International: Houston, TX, 1990.
[3]   Johnson, J.H., Kiepura, R.T. & Humpries, D.A. (eds.), *Corrosion*, ASM Handbook, Vol. 13, ASM International: Materials Park, OH, 1992.
[4]   *Dechema Corrosion Handbook*, Vols 1–12, (printed or CD ROM), Dechema, Elsevier, 1992–2003.
[5]   Shreir, L.L., Jarman, R.A. & Burstein, G.T., *Corrosion*, 3rd edn, Butterworth Heinemann: Oxford, UK, 1994.

[6] Trethewey, K.R. & Chamberlain, J., *Corrosion for Science and Engineering*, Longman Scientific &Technical: Singapore, 1995.

[7] Baboian, R. (ed.), *Corrosion Test and Standards Testing: Application and Interpretation*, ASTM Manual Series: MNL 20, ASTM: Philadelphia, 1995.

[8] Jones, D.A., *Principles and Prevention of Corrosion*, Prentice Hall: New Jersey, 1996.

[9] Mattsson, E., *Basic Corrosion Technology for Scientists and Engineers*, 2nd edn, The Institute of Materials: London, 1996.

[10] Sedriks, J.A., *Corrosion of Stainless Steels*, John Wiley & Sons, Inc.: New York, 1996.

[11] Baeckman, W., Schwenk, V. & Prinz, W. (eds.), *Handbook of Cathodic Protection*, 3rd edn, Elsevier Science: Burlington, MA, 1997.

[12] Munger, C.G., *Corrosion Prevention by Protective Coatings*, 2nd edn, NACE International: Houston, TX, 1999.

[13] Bogaerts, F.W. (ed.), *Active Library on Corrosion*, CD ROM ver. 2.0, Elsevier, 2000 (includes databases, corrosion atlas, books).

[14] Stansbury, E.E. & Buchanan, R.A., *Fundamentals of Electrochemical Corrosion*, ASM International: Materials Park, OH, 2000.

[15] Schütze, M. (ed.), *Corrosion and Environmental Degradation*, Vols 1 and 2, Wiley-VCH: Weinheim, 2000.

[16] Revie, W. R (ed.), *Uhlig's Corrosion Handbook*, John Wiley & Sons, Inc.: New York, 2000.

[17] Leygraf, Ch. & Graedel, T.E., *Atmospheric Corrosion*, Wiley Interscience: New York, 2000.

[18] Bard, A.J. & Faulkner, L.R., *Electrochemical Methods*, John Wiley & Sons, Inc.: New York, 2001.

[19] Marcus, P. & Oudar, J. (eds.), *Corrosion Mechanisms in Theory and Practice*, 2nd edn, Marcel Dekker: New York and Basel, 2002.

CHAPTER 3

Corrosivity of atmospheres – derivation and use of information

D. Knotkova & K. Kreislova
SVUOM Ltd., Prague, Czech Republic.

1 Introduction

Information on corrosivity is important for the selection of materials and protective systems for various products and objects. Most elaborated is the information on corrosivity of atmospheres in which the classification system and the methods of corrosivity derivation are standardized. In this chapter, ISO and EN standards are presented, the standards are compared and the solutions that will improve these standards are proposed. The monitoring methodology for indoor and outdoor atmospheric environments uses procedures that allow to carry out monitoring of objects (passive dosimetry, material sensors).

A comprehensive approach to the use of the information on corrosivity of atmospheres for proper selection of materials and anticorrosive measures is presented.

Data on corrosivity and trends of changes of corrosivity can be elaborated into maps of regional layout. The data can be also used for economic and noneconomic evaluation of corrosion damage.

2 Corrosivity as a property of material-and-environment system

Metals, protective coatings and other materials are subject to deterioration when exposed to atmospheric environments. Information on corrosivity is the basic requirement for the selection of materials, proposal of protective systems, estimation of service life and economic and noneconomic evaluations of damages caused by corrosion.

Materials deteriorate even in normal natural conditions. Various deterioration factors are decisive for individual types of materials. Pollution has a significant influence on the degradation of materials. The degradation process is accelerated by pollution from the moment of entering the environment, not after reaching critical levels. The response of materials differs according to their composition and other characteristics. A vast majority of materials exposed to atmospheric environments are sensitive to the impact of sulfurous substances, chloride aerosols and acidity of precipitation. Corrosion rate is strongly influenced by the concentration of pollution in the air. Most processes occur in the presence of humidity or directly under layers of electrolytes. The latest knowledge of atmospheric corrosion of metals leads to the conclusion that other components of atmospheric pollution are gaining importance in today's atmospheres where the content of sulfur dioxide has decreased. Recently, nitrogen dioxides and ozone have been reported to have negative impacts on some materials. Dust particles deposited on surfaces can also significantly increase corrosion attack of metals because they can contain corrosion stimulants, retain other impurities and can be a place for increased absorption of water. The degradation of materials in general, but specifically of metals, is caused by combined pollution.

However, the category of 'corrosivity' is related to the properties of the so-called corrosion system which can be very complex. Basically, this system involves:

- metal material (materials);
- atmospheric environment (characterized by temperature–wetness complex, or time of wetness and level of atmospheric pollution);
- technical parameters (construction design, shape and weight, manner of treatment, joints, etc.);
- conditions of operation.

The selection of metals, alloys or metallic coatings and the resistance of products to corrosion are influenced by the designed service life and the conditions of operation as well as the corrosivity of atmosphere.

While technical parameters and conditions of operation are highly specific for individual cases, interaction between materials and the environment can be evaluated on a more general level, and the evaluation process can be standardized. The degree of stress can be evaluated or expressed in stress classes that form the classification system of the corrosivity of atmospheres.

Best elaborated are the methods of derivation of corrosivity and classification systems for metals and their protective systems. A more general but rather indirect use of the knowledge is possible if the limiting factors of individual procedures are known. Corrosion engineering uses procedures that make use of a unified classification system for metals, metallic and other inorganic coatings and paints. A more general use, e.g. for assessment of stress on construction objects, is limited.

The methods of derivation and direct determination of corrosion rates are time-consuming and demand a lot of experiments. In principle, two approaches are available – either the initial corrosion rate is determined by direct one-year corrosion test allowing for qualified estimation of stabilized corrosion rate to follow or is calculated from the environmental data that are adequately available at present using 'deterioration equations' that are sometimes also called 'dose/response' (D/R) functions. However, the formulation of these equations is based on extensive corrosion tests carried out at a network of testing sites.

3 Classification of corrosivity for technical purposes – basis for standardization

Atmosphere is a corrosive environment, the impact of which affects products and objects of all kinds including objects of cultural heritage. A large portion of the damage caused by corrosion is attributed to atmospheric corrosion. Therefore, it is observed for a long term and the concerned knowledge has been generalized, which has gradually allowed for the elaboration of various international and national practical classification systems [1] for the evaluation of corrosivity of atmospheres. In the 1990s, these efforts resulted in a series of ISO standards that define not only a corrosivity classification system but also the procedures of its derivation and guiding corrosion values for long-term exposure of metals corresponding to individual corrosivity categories.

The standards for corrosivity classification were elaborated within the framework of ISO/TC 156/WG 4 *Atmospheric corrosion testing and classification of corrosivity of atmosphere*. The purpose was to elaborate standards that would meet the needs of corrosion engineers and users of technical products. Standards are well accepted and introduced by other technical committees (ISO/TC 107, ISO/TC 35/SC 14, CEN/TC 262, CEN/TC 240). Engineering application of the classification system is supported by the guidelines formulated in ISO 11303 *Corrosion of metals and alloys – Guidelines for selection of protection methods against atmospheric corrosion*.

Characterization of the standardized system
The standardized system is presented schematically in Fig. 1. It covers the following standards:

- ISO 9223:1992 Corrosion of metals and alloys – Corrosivity of atmospheres – Classification
- ISO 9224:1992 Corrosion of metals and alloys – Corrosivity of atmospheres – Standard values for corrosivity categories
- ISO 9225:1992 Corrosion of metals and alloys – Corrosivity of atmospheres – Measurement of pollution
- ISO 9226:1992 Corrosion of metals and alloys – Corrosivity of atmospheres – Determination of corrosion rate of standard specimens for the purpose of evaluating corrosivity.

A standardized classification system is a simple system with a consistent structure, which basically meets requirements for systems designed for wide

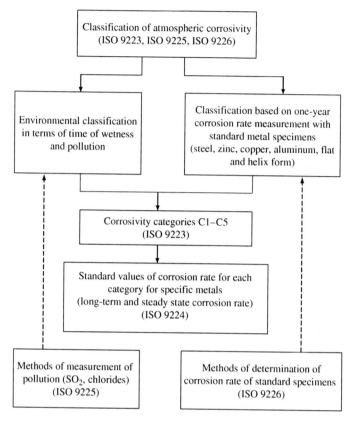

Figure 1: Scheme for classification of atmospheric corrosivity approach in ISO 9223–9226.

engineering use. The basic assumptions required for wide applicability of the system (Fig. 1) are:

- Possibility of a comparable derivation of corrosivity from annual averages of three decisive factors of the environment and from results of a one-year corrosion test.
- Possibility to associate long-term and steady-state corrosion rates with a corrosivity category derived from annual environmental or corrosion values.
- Low-cost and easy-to-use methods of measurements of pollution concentration or deposition rate.

Problems that are not fully solved:

- Standardized method for determination of one-year corrosion loss is not sensitive enough for determination of corrosion attack at C1 corrosivity category.
- The fact that the system does not provide corrosivity information for values over C5 corrosivity category does not mean that higher corrosion losses cannot be

Table 1: Classification of time of wetness (TOW), sulfur compounds based on sulfur dioxide (SO_2) concentration and airborne salinity contamination (Cl^-).

TOW	$h\,yr^{-1}$	SO_2	$\mu g\,m^{-3}$	$mg\,m^{-2}\,d^{-1}$	Cl	$mg\,m^{-2}\,d^{-1}$
τ_1	≤ 10	P_0	≤ 12	≤ 10	S_0	≤ 3
τ_2	10–250	P_1	12–40	10–35	S_1	3–60
τ_3	250–2500	P_2	40–90	35–80	S_2	60–300
τ_4	2500–5500	P_3	90–250	80–200	S_3	300–1500
τ_5	>5500					

reached, but this limitation must be explained within the standard so that the applicability of the standard is better defined.

3.1 Classification based on characterization of atmosphere

Three environmental parameters are used to assess corrosivity categories: time of wetness (TOW), sulfur compounds based on sulfur dioxide (SO_2) and airborne salinity contamination (Cl^-). For these parameters classification categories are defined as τ (TOW), P (SO_2) and S (Cl^-), based on measurements of the parameters (Table 1).

TOW is estimated from temperature–humidity (T-RH) complex as time when relative humidity is greater than 80% at a T-value greater than 0°C. TOW calculated by this method does not necessarily correspond to the actual time of exposure to wetness because wetness is influenced by many other factors. However, for classification purposes the calculation procedure is usually accurate.

Given the categories for TOW, SO_2 and Cl as defined in Table 1, the standard provides different look-up tables for carbon steel, zinc, copper and aluminum, which allows for estimation of corrosivity categories C1–C5.

3.2 Classification based on measurement of corrosion rate

Table 2 shows corrosivity categories for different metals. It needs to be stressed that the highest category (C5) has specified upper limits. Values above these limits have been detected in the field but they are outside the scope of the present standard.

The system provided by ISO 9223–9226 standards is used worldwide, although a few weak points of this system based on generalized knowledge of atmospheric corrosion of metals and results of many partial site test programs from around 1990 are known.

When a similar classification standard EN [2] was introduced later on, this fact was reflected in the differentiation of methods used to derive corrosivity from results of one-year exposure of standard specimens – same as in ISO 9223 and ISO 9226 – and to estimate corrosivity from atmospheric environmental data.

Table 2: ISO 9223 Corrosivity categories for carbon steel zinc, copper and aluminum based on corrosion rates.

Corrosivity	Category	Carbon steel (μm yr^{-1})	Zinc (μm yr^{-1})	Copper (μm yr^{-1})	Aluminum (g m^{-2} yr^{-1})
Very low	C1	≤ 1.3	≤ 0.1	≤ 0.1	Negligible
Low	C2	1.3–25	0.1–0.7	0.1–0.6	≤ 0.6
Medium	C3	25–50	0.7–2.1	0.6–1.3	0.6–2
High	C4	50–80	2.1–4.2	1.3–2.8	2–5
Very high	C5	80–200	4.2–8.4	2.8–5.6	5–10

The same applies, although not so clearly, to the classification system for coatings elaborated for ISO 12944-2. The system is also based on the classification defined in Table 2.

Adjustment of atmospheric corrosivity classification systems has been going on in ISO/TC 156/WG 4 *Corrosion of metals and alloys – Atmospheric corrosion testing and atmospheric corrosivity classification* during recent years mainly with the cooperation of Czech, Swedish, American, Russian, Spanish and Australian members. An important contribution to the improvement of the corrosivity classification system was systematic evaluation and adjustment, including statistical processing of the ISO CORRAG program [3, 4]. This program was carried out at more than 50 sites in 13 countries and was designed to follow the ISO classification system both in its methodology and structure. Later on, the MICAT program in Spain, Portugal and Iberoamerican countries [5] was organized in a similar manner.

Based on the activities following the Convention on Long-Range Transboundary Air Pollution in agreement with the *UN ECE International Co-operative Programme on Effects on Materials Including Historic and Cultural Monuments*, the influence of pollution on the corrosion rate of basic construction metals has been systematically observed for a long term, and deterioration equations that express dependence of the effect on environmental factors (RH, SO_2, Cl^-, O_3, temperature, acidity of precipitation) have been derived for main materials. These functions, which express the dependence of the degree of deterioration on pollution and other parameters, can be transformed into service-life functions in most cases that can be used for an economic evaluation of the damage. The derivation of deterioration functions that could be widely applied will be the basic contribution to the improvement of the classification system in the future.

The classification systems presented can be fully used to evaluate environmental stress, the selection of anticorrosion measures or estimations of service life according to the provisions of individual standards. The selection of accelerated corrosion tests that verify proposals of anticorrosion measures for given environments (i.e. selection of type of tests and degree of severity of tests) in relation to the expected corrosivity of the atmosphere in the exposure conditions is not expressed directly which is justified because accelerated corrosion tests do not correlate with corrosion stress in the same manner as in the case of corrosion stress in conditions of real exposure.

4 Classification of low corrosivity of indoor atmospheres

Procedures prescribed in ISO 9223 provide reasonably good results in the derivation of corrosivity categories for outdoor atmospheric environments. This classification system is too rough for indoor environments of low corrosivity. More sensitive procedures have to be chosen for the derivation of corrosivity categories in less aggressive indoor environments, such as places where electronic devices, hi-tech technical products or works of art and historical objects are stored. A new system for the classification of corrosivity of indoor atmospheres has been developed and the following three documents were well accepted:

- ISO 11844 Part 1: Corrosion of metals and alloys – Classification of low corrosivity of indoor atmospheres – Determination and estimation of indoor corrosivity,
- ISO 11844 Part 2: Corrosion of metals and alloys – Classification of low corrosivity of indoor atmospheres – Determination of corrosion attack in indoor atmospheres,
- ISO 11844 Part 3: Corrosion of metals and alloys – Classification of low corrosivity of indoor atmospheres – Measurement of environmental parameters affecting indoor corrosivity.

The aim of newly elaborated standards is to provide a consistent method of low indoor corrosivity classification with improved procedures for determination of indoor corrosivity categories based on evaluation of corrosion attack and estimation of indoor corrosivity categories based on environmental characterization. The scheme representing the approach for low indoor corrosivity classification is presented in Fig. 2.

The new standards are based on the results from systematic investigations of parameters affecting corrosion performed both in climate chambers and extensive field exposures in widely differing types of indoor environments [6]. Some characteristics of indoor atmospheric environments in relation to corrosion of metals are represented by the following:

- Temperature, humidity and their changes cannot be derived directly from outdoor conditions. They depend on the purpose of use of an indoor space in unconditioned atmospheric environments.
- Transfer of outdoor pollution depends on the manner and degree of shelteredness or on controlled conditions in indoor atmospheric environment (filtration, conditioning).
- Successive accumulation of particles and increasing conductivity of deposited water extracts can change corrosivity for longer indoor exposures.

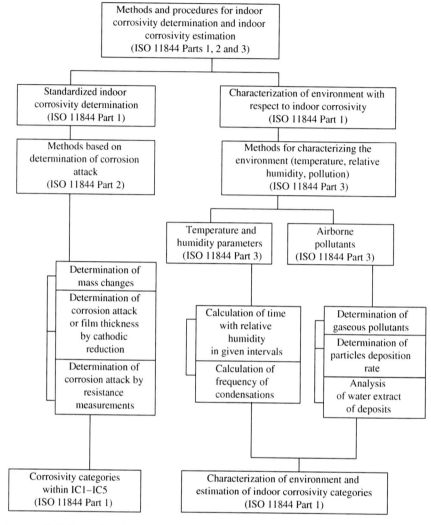

Figure 2: Scheme for classification of low corrosivity in indoor atmospheres in ISO 11844 Parts 1–3.

- Besides emissions from outside objects, release from building and construction materials, and human presence and activity can represent sources of specific polluting substances.

Various sensitive techniques for the determination of corrosion attack and different methods for monitoring and measurement of environmental parameters [6] were used during investigations. The main aspects of the new approach are as follows:

- Determination of corrosivity categories of indoor atmospheres is based on the assessment of corrosion attack on standard specimens of four metals. The choice

Table 3: Classification of corrosivity of indoor atmospheres based on corrosion rate measurements by mass loss determination of standard specimens.

Corrosivity category	Corrosion rate (r_{corr}) per unit of surface (mg m^{-2} yr^{-1})	
	Copper	Silver
IC 1 Very low indoor	$r_{corr} \leq 50$	$r_{corr} \leq 170$
IC 2 Low indoor	$50 < r_{corr} \leq 200$	$170 < r_{corr} \leq 670$
IC 3 Medium indoor	$200 < r_{corr} \leq 900$	$670 < r_{corr} \leq 3000$
IC 4 High indoor	$900 < r_{corr} \leq 2000$	$3000 < r_{corr} \leq 6700$
IC 5 Very high indoor	$2000 < r_{corr} \leq 5000$	$6700 < r_{corr} \leq 16\,700$

of metals used as standard specimens has changed taking the importance of occurrence in considered environments into account (silver replaced aluminum).

• Sensitive evaluation techniques have been used for the determination of mass changes using ultra-microbalance, for the determination of the thickness of corrosion products by electrolytic cathodic reduction, or for the determination of resistance changes in thin film resistance sensors. Microbalance based on quartz crystal is used for evaluation of corrosive effects of low-corrosivity indoor atmospheres mainly in Sweden and USA. This method is expected to be involved in ISO 11844-2 standard in addition to procedures already standardized.

• It is known that SO_2 and chloride levels do not have a dominating influence on corrosion rate in present-day indoor atmospheres. Instead a multipollutant effect of various gaseous pollutants including organic acids and aldehydes and particulate matters, in combination with temperature-humidity complex, is of decisive importance.

The proposed standards for corrosivity in indoor atmospheres have a similar structure and are based on the same principles as the present standards ISO 9223–9226. However, they cover less corrosive atmospheres corresponding to corrosivity categories C1 and C2. Even if the methods for the determination of corrosivity are far more sensitive and the characterization of the environment takes specific features of low-corrosivity environments into account, the use of common reference materials (copper, zinc, carbon steel) makes the two sets of standards comparable. Table 3 gives an example of indoor classification based on the determination of corrosion rates in low-corrosivity environments for copper and silver.

5 Derivation of corrosivity of exposure locations, monitoring

5.1 Estimation of degree of stress caused by pollution and other environmental factors on materials, products and constructions

Preliminary assessment of stress caused by the environment on materials and objects can be done by estimating corrosivity category by finding a standardized environment in the classification system same as the environment to which an object is exposed.

The estimation is based on the knowledge of main climatic characteristics and level of pollution (average values). Published data can be used in most cases. This method is in compliance with standard EN 12500 *Protection of metal materials against corrosion – Corrosion likelihood in atmospheric environment – Classification, determination and estimation of corrosivity of atmospheric environment.*

The estimation of corrosivity category may lead to inexact interpretations; however, it should precede active monitoring. During the estimation, a description of an environment is compared to the standard environment; therefore, not all facts defining the corrosion system are included. The concentration of sulfur dioxide (SO_2) is expressed as the annual average value. Extreme effects of chlorides, which are typical for marine splash or very heavy salt spray, are beyond the scope of this classification system. Standard environments are presented in the informative part of the quoted standard (see Table 4).

It is easier to estimate the corrosivity of outdoor environments than of indoor environments with low corrosivity owing to the easier accessibility to and completeness of environmental data and deeper as well as the more complex knowledge of the mechanism and kinetics of corrosion in outdoor atmospheric environment. Therefore, more sensitive procedures for the determination and estimation of corrosivity category for low-corrosivity indoor environment are being gradually introduced on an international scale. The table of standard indoor environments presented corresponds with the ISO 11844-1 *Corrosion of metals and alloys – Classification of low corrosivity of indoor atmospheres – Part 1: Determination and estimation of indoor corrosivity* and lays down conditions for the estimation of corrosivity category of low-corrosivity indoor environments approximately between C1 and C2 categories in accordance with ISO 9223 standard.

Corrosivity categories IC1–IC5 (indoor corrosivity) shown in Table 5 basically correspond with the range of C1–C2 categories in Table 4. Metals and metallic coatings perform specific corrosion behavior in indoor atmospheres.

It is recommended that at least orientation measurements of temperature and relative humidity for estimations of corrosivity categories of indoor environments be carried out if these values are not already being observed. Measurements should involve at least one month of each typical climatic season.

The categorization of levels of average relative humidity in accordance with ISO 11844-1 is presented in Table 6.

Measuring pollution in indoor environments is rather difficult and is not carried out for the purpose of estimating the corrosivity category. However, the possibility of outdoor pollution penetrating into the indoor environment and formation of specific, especially organic, pollution needs to be considered. An important component of indoor environmental pollution is pollution with particles, namely particles containing ionogeneous fractions.

Table 4: Description of typical atmospheric environments related to estimation of corrosivity categories according to EN 12500.

Corrosivity category	Corrosivity	Standard environments (examples)	
		Indoor	Outdoor
C1	Very low	Heated spaces, low relative humidity and insignificant pollution, e.g. offices, schools, museums.	Dry or cold zone, atmospheric environment, very low pollution and time of wetness, e.g. certain deserts, central Antarctica.
C2	Low	Unheated spaces, varying temperature and relative humidity. Low frequency of condensation and low pollution, e.g. storage rooms, sports halls.	Temperate zone, atmospheric environment, low pollution ($SO_2 < 12\,\mu g\,m^{-3}$), e.g. rural areas, small towns. Dry or cold zone, atmospheric environment with short time of wetness, e.g. deserts, sub-arctic areas.
C3	Medium	Spaces with moderate frequency of condensation and moderate pollution from industrial production, e.g. food processing, plants, laundries, breweries, dairies.	Temperate zone, atmospheric environment with medium pollution (SO_2: $12–40\,\mu g\,m^{-3}$) or certain effect of chlorides, e.g. urban areas, coastal areas with low deposition of chlorides. Tropical zone, atmosphere with low pollution.
C4	High	Spaces with high frequency of condensation and high pollution from industrial production, e.g. industrial processing plants, swimming pools.	Temperate zone, atmospheric environment with high pollution (SO_2: $40–80\,\mu g\,m^{-3}$) or substantial effect of chlorides, e.g. polluted urban areas, industrial areas, coastal areas without spray of salt water, strong effect of decaying salts. Tropical zone, atmosphere with medium pollution.
C5	Very high	Spaces with almost permanent condensation and/or with high pollution from industrial production, e.g. mines, caverns for industrial purposes, unventilated sheds in humid tropical zones.	Temperate zone, atmospheric environment with very high pollution (SO_2: $80–250\,\mu g\,m^{-3}$) and/or strong effect of chlorides, e.g. industrial areas, coastal and off shore areas with salt spray. Tropical zone, atmosphere with high pollution and/or strong effect of chlorides.

Table 5: Description of typical environments related to estimation of indoor corrosivity categories according to ISO 11844-1.

Corrosivity category	Corrosivity	Standard environments	
		Heated spaces	Unheated spaces
IC1	Very low indoor	Controlled stable relative humidity (<40%), no risk of condensation, low levels of pollutants, no specific pollutants, e.g. computer rooms, museums with controlled environment.	With dehumidification, low levels of indoor pollution, no specific pollutants, e.g. military storages for equipment.
IC2	Low indoor	Low relative humidity (<50%), certain fluctuation of relative humidity, no risk of condensation, low levels of pollution, no specific pollutants, e.g. museums, control rooms.	Only temperature and humidity changes, no risk of condensation, low levels of pollution without specific pollutants, e.g. storage rooms with low frequency of temperature changes.
IC3	Medium indoor	Risk of fluctuation of temperature and humidity, medium levels of pollution, certain risk for specific pollutants, e.g. switch boards in power industry.	Elevated relative humidity (<50–70 %), periodic fluctuation of relative humidity, no risk of condensation, elevated levels of pollution, low risk of specific pollutants, e.g. churches in nonpolluted areas, outdoor telecommunication boxes in rural areas.
IC4	High indoor	Fluctuation of humidity and temperature, elevated levels of pollution including specific pollutants, e.g. electrical service rooms in industrial plants.	High relative humidity (up to 70%), risk of condensation, medium levels of pollution, possible effect of specific pollutants, e.g. churches in polluted areas, outdoor boxes for telecommunication in polluted areas.
IC5	Very high indoor	Limited influence of relative humidity, higher levels of pollution including specific pollutants like H_2S, e.g. electrical service rooms, cross-connection rooms in industries without efficient pollution control.	High relative humidity and risk of condensation, medium and higher levels of pollution, e.g. storage rooms in basements in polluted areas.

Table 6: Levels of average relative humidity [6].

Level	Relative humidity average (%)	Range of changes (%)
I	RH < 40	<10
II	40 ≤ RH < 50	10 ≈ 20
III	50 ≤ RH < 70	10 ≈ 30–40
IV	RH ≥ 70	<40

5.2 Monitoring effect of pollution and other environmental factors on materials and objects

Outdoor environment has negative effects on materials and objects; they are gradually damaged or even destroyed. The effects of individual environmental components are not separate but combined, though in some cases they affect in synergy.

The most complete data on environmental effects on materials are provided by results of systematic monitoring and their evaluation aimed at estimating corrosivity.

The approach to the proposal of monitoring effects of atmospheric environment is based on the provisions of standards ISO 9223–9226 and ISO 11844, Parts 1–3. Although measurements of selected characteristics of atmospheric environment are a part of monitoring, the most important part is the determination of effects on standard material specimens, which allows for the determination of the corrosivity categories of atmospheres based on the provision of quoted standards. Corrosion effects caused in general indoor environments are significantly smaller than effects caused in outdoor environments. Therefore, procedures for the determination of corrosion effects are differentiated so that desired sensitivity is reached.

5.2.1 Monitoring in outdoor atmospheric environments

Various monitoring units respecting basic provisions of ISO 8565 *Metals and alloys – Atmospheric corrosion testing – General requirements for field tests* were designed for monitoring in outdoor environments. Examples of such units are shown on Figs 3 and 4. These were produced and used for monitoring on St. Vitus Cathedral in Prague. The first unit is very simple; the design allows for acquiring experimental results in accordance with ISO 9223 standard, which means that standard metal specimens are exposed for one year and the concentration of basic gaseous pollutions is determined using passive samplers. The deposition of NaCl has not been considered because it is not important within the central area of Prague. The design of the second monitoring unit is more complicated. Environmental effects can be observed on a wider variety of materials including natural stone; pollution deposition rate on selected surfaces is observed; and the activity of a microbial component is assessed. Datalogers observing temperature and relative humidity are placed very well too.

Figure 3: Monitoring unit including basic materials and environmental sensors.

Figure 4: Rack with monitoring system.

Passive dosimetry is the most frequently used method to measure gaseous pollutions, which is most suitable when placing monitoring units in the field or on objects. Other methods or commercial devices can be used for measurement of gaseous pollution as well.

Newly proposed monitoring systems also include sensors to observe the deposition of dust fallout; the collected fallout is then analyzed for the presence of fractions stimulating corrosion. If monitoring units are installed on important, e.g. historic buildings, darkening of surfaces (so-called soiling) of selected materials is observed too.

The observation of environmental effects on various materials extends the validity of the original information. Most experience was achieved in case of exposure of material sensors on the basis of natural stone. Procedures were elaborated in UK, Germany and Czech Republic [7–9]. Plates of selected types of limestone and sandstone are exposed. Exposure methods and evaluation procedures are presented for example in ref. [10]. Weight changes and content of sulfate, nitrate and chloride ions in surface layers after exposure are evaluated, and SEM-EDS analysis of surfaces is carried out.

Certain types of glass are specifically sensitive to effects of atmospheric pollution, which provided a basis for proposals of glass monitoring sensors [11]. Use of these sensors for monitoring corrosivity of environment is included in the national standard [12].

Derivation of corrosivity for coatings in accordance with ISO 12944-2 standard can be based on monitoring carried out on metals. Specific knowledge of protective effectiveness is then provided by tests on samples of organic coatings on steel with cuts.

5.2.2 Monitoring for indoor atmospheric environments

Monitoring in indoor environments is based also mainly on the determination of effects caused by the environmental impact on materials or specially designed material sensors. Special conditions of air flow and their influence on the spread of corrosivity must be considered when placing monitoring units.

No special monitoring units are usually designed for monitoring in indoor spaces of objects, especially historical ones. Small elements can be used for monitoring. They are placed on suitable horizontal surfaces.

The method of determination of quantitative corrosion changes – weight gain or loss – on standard metal specimens is very expensive and material, demanding on time and equipment, and is used as a part of monitoring in indoor environments in special cases only. The determination of corrosion attack by cathodic reduction of formed layers of corrosion products on copper or silver is also not used normally; only selected testing laboratories use this method.

Exposure of atmospheric resistance corrosion sensors and their evaluation is very easy but costs of sensors and scanning units are not low. High Sensitivity Atmospheric Corrosion Sensors – model 610 made by Rohrback Cobasco System INC. are an example of sensors used.

Considering the amount of work and costs required in most quantitative methods of determining low-corrosion effects in indoor atmospheric environments, the expected effects can be evaluated within the monitoring process by qualitative changes on surfaces of copper and silver samples. Specially prepared pieces of copper and silver or copper with silver coating are used, and changes are evaluated by appearance. Changes can be expressed in codes or degrees. First, the appearance of sample surfaces with a layer of deposits is evaluated, and then, the appearance is evaluated after deposits are removed with a fine brush. The following aspects are evaluated:

- general character of changes caused;
- character of a layer of deposits on a sample surface and thickness of the layer;
- corrosion attack caused by gaseous pollutants;
- character of a sample surface after a layer of deposits is removed.

A method of semi-quantitative evaluation of corrosion changes caused on copper and silver samples was proposed and verified in the Czech Republic. All samples can be digitally scanned in the same way after exposure. The electronic form of pictures allows for evaluation using computer analysis of pictures, which provides for an objective evaluation of surface changes. Brightness cuts were made from digitized pictures of the exposed samples. These located and quantified (up to certain extent) the appearance evaluation carried out.

The principles for the selection of methods for measuring pollution in indoor atmospheric environments are presented in ISO 11844-3. Again, the method of passive dosimetry is recommended for use preferentially. The occurrence of organic components of atmospheric pollution (formaldehyde, organic acids), ammonium components and hydrogen sulfide need to be observed in areas visited by general public. The evaluation of fallout of particles and the determination of aggressive fractions of fallout complement characterization of indoor environments in relation to their corrosivity.

6 Using new knowledge to improve classification systems

The fundamental contribution to the adjustment of the existing corrosivity classification system is evaluation and statistical treatment of data provided by the international exposure programs ISOCORRAG and MICAT completed by specific data from tropical and cold regions.

A huge effort was made for a complex elaboration of results by working groups carrying out international testing programs aimed especially at deepening the knowledge of the effect of individual environmental factors and groups of these factors on kinetics of the corrosion process. Results are elaborated in reports and publications [13–15].

The basic contributions are proposals of new and better-elaborated dose–response functions that are generally valid for a wide regional scale. A gradual introduction of these functions to the standardized classification system will remove or at least significantly lower the most often criticized disproportion in the derivation of corrosivity of atmospheres from results of corrosion tests and from the knowledge of environmental data.

Although the classification system will significantly improve after new deterioration functions are introduced and will be easier to use too, both methods cannot be fully equivalent for reasons of principle.

Direct corrosion test covers the whole complex of environmental stress during one specific year of exposure, but the average stress during the expected service life of an object may differ. Using the D/R function, the average values of environmental factors valid for given time periods and regions can be substituted. However, functions involve only selected decisive environmental factors and calculated value of corrosion loss must be considered as a point determination of the value, which shows deviations that can be statistically determined.

The dose–response functions are all based on data after one year of exposure and can, therefore, only be used for classification purposes and not for assessing service lives of materials in different environments.

The dose–response functions for four standard metals are given describing the corrosion attack after one year of exposure in open air as a function of SO_2 dry deposition, chloride dry deposition, temperature and relative humidity. The functions are based on the results of large corrosion field exposures and cover climatic earth conditions and pollution situations in the scope of ISO 9223.

The dose–response functions for calculation of the yearly corrosion loss of structural metals are given below.

Carbon steel

$$r_{corr} = 1.77[SO_2]^{0.52} \exp(0.020RH + f_{St}) + 0.102[Cl^-]^{0.62} \exp(0.033RH + 0.040T)$$

$f_{St} = 0.150(T - 10)$ when $T \leq 10°C$, otherwise $-0.054(T - 10); N = 128, R^2 = 0.85$.

Zinc

$$r_{corr} = 0.0129[SO_2]^{0.44} \exp(0.046RH + f_{Zn}) + 0.0175[Cl^-]^{0.57} \exp(0.008RH + 0.085T)$$

$f_{Zn} = 0.038(T - 10)$ when $T \leq 10°C$, otherwise $-0.071(T - 10); N = 114, R^2 = 0.78$.

Copper

$$r_{corr} = 0.0053[SO_2]^{0.26} \exp(0.059RH + f_{Cu}) + 0.0125[Cl^-]^{0.27} \exp(0.036RH + 0.049T)$$

$f_{Cu} = 0.126(T - 10)$ when $T \leq 10°C$, otherwise $-0.080(T - 10); N = 121, R^2 = 0.88$.

Aluminum

$$r_{corr} = 0.0042[SO_2]^{0.73} \exp(0.025RH + f_{Al}) + 0.0018[Cl^-]^{0.60} \exp(0.020RH + 0.094T)$$

$f_{Al} = 0.009(T - 10)$ when $T \leq 10°C$, otherwise $-0.043(T - 10); N = 113, R^2 = 0.65$.

In the above equations, r_{corr} is the first-year corrosion rate of the metal in μm/year. For explanation of the remaining parameters see Table 7, which also gives the measured intervals of the parameters.

Care should be taken when extrapolating the equations outside the intervals of environmental parameters for their calculation (e.g. in coastal environments).

The R^2 values are between 0.8 and 0.9, except for aluminum, where it is substantially lower. Aluminum experiences localized corrosion but corrosion attack is calculated as uniform corrosion, which might be one reason for the low value.

Table 7: Parameters used in dose–response functions including symbol, description, interval measured in the program and unit. All parameters are expressed as annual averages.

Symbol	Description	Interval	Unit
T	Temperature	−17.1–28.7	°C
RH	Relative humidity	34–93	%
SO$_2$	SO$_2$ deposition	0.7–150.4	mg m^{-2} day^{-1}
Cl	Cl$^-$ deposition	0.4–760.5	mg m^{-2} day^{-1}

If these functions are to be used, the inaccuracies of the calculated value must be statistically defined in a professional manner. Models of dose–response functions do not consider two selection errors:

- selection error of regression model coefficients, i.e. selection error of the whole regression function;
- residual spread of individual observations around the regression function.

Characteristics of the regression function are also influenced by reliability of measurement of the input quantities:

- Errors have been estimated for weathering steel, zinc, copper, bronze and aluminum both in case of determining corrosion attack from measurements and in case of calculating corrosion attack from ICP Materials dose–response functions.
- Measurement error is generally ±5% or lower.
- Calculation error is significantly higher than the measurement error, about ±30%, mostly due to uncertainty in the dose–response function and to a lesser extent due to uncertainty in environmental parameters, which is around ±7% on average if dry deposition prevails.

7 Trends in corrosivity of atmosphere within Europe

Results of global efforts to decrease industrial emissions can be evaluated by gradual decrease of corrosivity at systematically observed locations. Decrease of corrosivity has been going on since 1985 in Europe, and effectiveness of measures taken on the decrease of corrosion effects has been observed since 1990.

As a part of the Task Force of *UN ECE ICP Effects on materials including historic and cultural monuments* the program subcenter, SVUOM (Czech Republic) evaluates development of corrosivity according to the values of annual corrosion loss of carbon steel and zinc which are regularly determined at a site network [16, 17]. These two metals are basic standard materials for the derivation of corrosivity in the classification system according to ISO 9223.

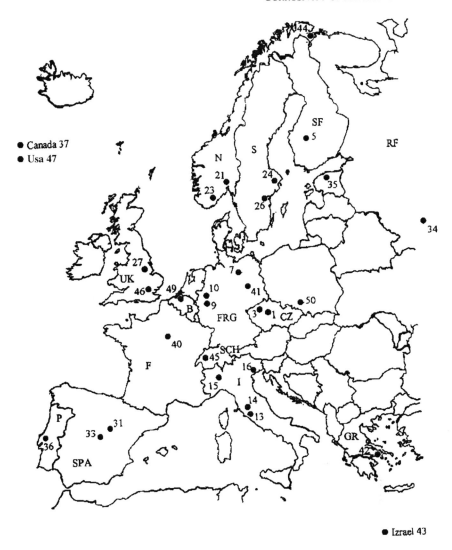

Figure 5: Map showing location of test sites.

Sites cover the area of Central and Western Europe sufficiently, and two sites are located in Eastern Europe (Fig. 5). Two sites located outside Europe provide at least partial comparison in a wider regional sense.

Samples of unalloyed carbon steel and zinc were exposed during the program as standard specimens for the determination of corrosivity of test sites (according to ISO 9226). Within the framework of the program the exposures of these samples were repeated to validate decreasing trend of corrosivity due to decrease of pollution by SO_2. Comparisons of corrosion losses for different exposure periods are given in

Table 8: Comparison of corrosion losses (g m^{-2}) of unalloyed steel after one-year exposure in open atmosphere in six exposure periods.

Test site	Period					
	1987–88	1992–93	1994–95	1996–97	1997–98	2000–01
1 Prague	438 C4	271 C3	241 C3	232 C3	182 C2	138 C2
3 Kopisty	557 C4	350 C3	352 C3	293 C3	239 C3	224 C3
5 Ahtari	132 C2	48 C2	59 C2	54 C2	53 C2	51 C2
7 Waldhof-L.	264 C3	231 C3	166 C2	156 C2	144 C2	148 C2
9 Langenfeld	293 C3	231 C3	210 C3	206 C3	204 C3	101 C2
10 Bottrop	373 C3	347 C3	294 C3	296 C3	311 C3	293 C3
13 Rome	178 C2	–	124 C2	109 C2	134 C2	80 C2
14 Casaccia	235 C3	–	148 C2	135 C2	125 C2	98 C2
15 Milan	366 C3	–	198 C2	149 C2	173 C2	184 C2
16 Venice	245 C3	–	212 C3	188 C2	211 C3	149 C2
21 Oslo	229 C3	135 C2	101 C2	99 C2	93 C2	97 C2
23 Birkenes	194 C2	132 C2	109 C2	114 C2	101 C2	114 C2
24 Stockholm S.	264 C3	120 C2	103 C2	104 C2	125 C2	117 C2
26 Aspvreten	147 C2	75 C2	81 C2	69 C2	62 C2	69 C2
27 Lincoln Cath.	315 C3	308 C3	237 C3	–	270 C3	195 C2
31 Madrid	222 C3	162 C2	151 C2	159 C2	72 C2	77 C2
33 Toledo	45 C2	26 C2	36 C2	36 C2	54 C2	47 C2
34 Moskow	181 C2	141 C2	121 C2	123 C2	135 C2	136 C2
35 Lahemaa	185 C2	–	–	–	106 C2	95 C2
36 Lisbon	224 C3	308 C3	204 C3	289 C3	214 C3	226 C3
37 Dorset	149 C2	110 C2	104 C2	120 C2	116 C2	99 C2
40 Paris	–	–	–	–	137 C2	143 C2
41 Berlin	–	–	–	174 C2	179 C2	174 C2
43 Tel Aviv	–	–	–	–	324 C3	256 C3
44 Svanvik	–	–	–	160 C2	166 C2	148 C2
45 Chaumont	–	–	–	94 C2	67 C2	58 C2
46 London	–	–	–	–	177 C2	170 C2
47 Los Angeles	–	–	–	–	136 C2	159 C2
49 Antverps	–	–	–	–	171 C2	186 C2
50 Katowice	–	–	–	–	–	271 C3

Tables 8 and 9 [17]. There are also included relevant corrosivity categories according to ISO 9223.

Corrosion losses of both metals have decreased significantly. The decreasing trend is more evident in the case of carbon steel than of zinc. Corrosion of carbon steel is more affected by SO_2 pollution; corrosion of zinc is more affected by climatic conditions, especially by wetness of sample surfaces.

Table 9: Comparison of corrosion losses (g m^{-2}) of zinc after one-year exposure in open atmosphere in five exposure periods.

Test site	Period				
	1989–90	1992–93	1994–95	1996–97	2000–01
1 Prague	7.0 C3	7.7 C3	5.6 C3	5.6 C3	3.9 C2
3 Kopisty	11.5 C3	11.6 C3	12.1 C3	8.8 C3	4.6 C2
5 Ahtari	7.6 C3	6.6 C3	4.6 C2	3.1 C2	5.8 C3
7 Waldhof-L.	7.8 C3	9.0 C3	4.2 C2	3.5 C2	4.6 C2
9 Langenfeld	6.6 C3	9.0 C3	7.6 C3	5.7 C3	5.8 C3
10 Bottrop	10.6 C3	15.2 C3	7.8 C3	8.6 C3	9.2 C3
13 Rome	9.7 C3	–	3.4 C2	3.8 C2	3.1 C2
14 Casaccia	9.9 C3	–	3.1 C2	3.1 C2	2.5 C2
15 Milan	12.1 C3	–	5.5 C3	4.4 C2	5.2 C3
16 Venice	7.6 C3	–	6.1 C3	4.1 C2	4.6 C2
21 Oslo	5.6 C3	6.6 C3	3.5 C2	2.3 C2	3.5 C2
23 Birkenes	8.4 C3	10.5 C3	5.0 C3	3.5 C2	8.5 C3
24 Stockholm S.	6.0 C3	4.5 C2	4.2 C2	3.2 C2	4.4 C2
26 Aspvreten	6.7 C3	4.8 C2	6.0 C3	2.6 C2	3.5 C2
27 Lincoln Cath.	12.3 C3	10.6 C3	7.0 C3	–	5.2 C3
31 Madrid	4.8 C2	3.5 C2	2.3 C2	2.4 C2	2.1 C2
33 Toledo	3.9 C2	4.7 C2	1.7 C2	2.2 C2	2.3 C2
34 Moskow	8.6 C3	6.5 C3	4.6 C2	4.1 C2	6.8 C3
35 Lahemaa	–	–	–	–	5.1 C3
36 Lisbon	–	10.4 C3	5.6 C3	4.1 C2	8.0 C3
37 Dorset	6.2 C3	5.2 C3	6.1 C3	2.6 C2	3.1 C2
40 Paris	–	–	–	–	4.9 C2
41 Berlin	–	–	–	6.3 C3	5.8 C3
43 Tel Aviv	–	–	–	–	7.6 C3
44 Svanvik	–	–	–	2.8 C2	3.9 C2
45 Chaumont	–	–	–	3.0 C2	5.0 C2
46 London	–	–	–	–	6.1 C3
47 Los Angeles	–	–	–	–	5.0 C2
49 Antverps	–	–	–	–	8.2 C3
50 Katowice	–	–	–	–	9.9 C3

The decrease of corrosion rate is more significant for test sites reaching high levels of pollution by SO_2 in the period 1987–88. Pollution situation on test site No. 36 in Lisbon was influenced by periodical restoration works on the building where the test site is located.

The corrosion loss of carbon steel was reduced during the period 1987–2001 on 20 test sites in Europe by 56% from the average value of 250 g m^{-2} to 140 g m^{-2} (Fig. 6).

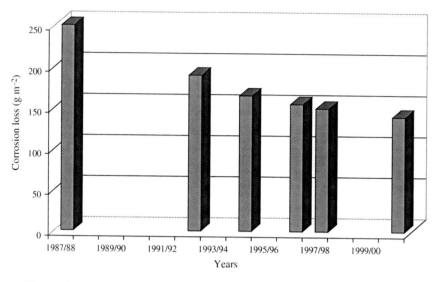

Figure 6: Average corrosion loss of carbon steel on 20 test sites in Europe.

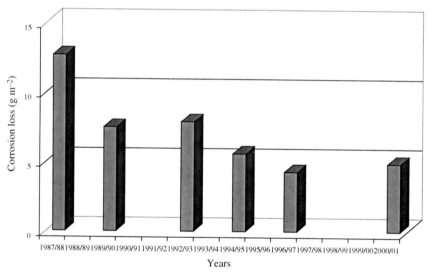

Figure 7: Average corrosion loss of zinc at 20 test sites in Europe.

This decrease in corrosion loss has been caused by decreasing concentration of SO_2 in the air together with decreasing acidity of precipitation.

The most significant decrease of SO_2 pollution along with decrease of corrosion rates occurred towards the end of 1980s. During the last five years pollution by SO_2 has not been changing so dramatically at the majority of test sites as it did in the previous decade. Yearly corrosion loss of carbon steel in periods 1997–98 and 2000–01 has been stable, reaching similar levels at majority of test sites as in the original program. The decreasing trend of corrosion loss of carbon steel will continue in the future, but probably the rate will be slower.

Yearly corrosion loss of zinc was reduced during the period 1987–2001 at 20 test sites in Europe by 61% from the average value of 13 g m^{-2} to 5 g m^{-2} (Fig. 7). The decrease was caused by decreasing concentration of SO_2 in the air together with decreasing acidity of precipitation.

The corrosion losses of carbon steel and zinc evaluated in 2003 confirm the trends described above.

More detailed data on development of environmental situation and decrease of corrosivity at two Czech sites involved in the quoted program are shown in Figs 8 and 9 where environmental data (especially SO_2 concentration) for all exposure periods are available.

Compared to other sites involved in UN ECE ICP program these two sites had higher levels of industrial pollution, therefore, decrease of corrosivity is significant there. The atmospheric test site Kopisty is located in a part of the so-called Black Triangle – an industrial area in Northern Bohemia. Determined values prove that measures that were taken were effective even in such extreme conditions.

A significant decrease in SO_2 concentration was reported also in Germany – a country with industrial regions of high corrosivity. In terms of annual values, SO_2 air concentration decreased by 88% between 1985 and 1998 (Fig. 10) [18]. The reduction in the concentration of SO_2 in the air clearly reflects a decreasing tendency of emissions in Germany as well as Europe as a whole. The reasons for the reduction of emissions result from efforts of emission abatement and political and economic changes that started towards the end of 1980s in Eastern Germany and East European countries.

8 Technical use of information on corrosivity

Technical use of information on corrosivity consists mainly in rationalizing selection of anticorrosion measures. Assessing atmospheric environment with respect to atmospheric corrosivity is a premise for further economic evaluation of the corrosion damage caused.

However, should planned service life of an object or anticorrosion measure be reached, other aspects need to be considered too. Rational engineering approach to selection of materials and anticorrosion measures in relation to corrosivity of exposure environment and designed service life is determined by the provisions of newly introduced ISO 11303 *Corrosion of metals and alloys – Guidelines for selection of protection methods against atmospheric corrosion.*

Generally, protection against atmospheric corrosion can be achieved by selecting suitable material and design of a product with respect to protection against corrosion, by reducing corrosivity of the environment and by covering a product with appropriate protective coatings.

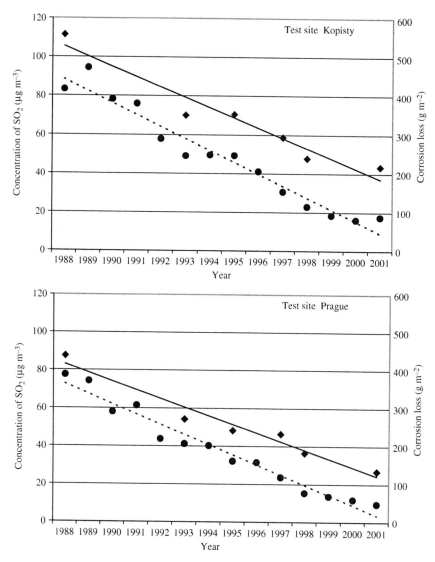

Figure 8: Trend of decrease of SO_2 concentration ($\mu g\,m^{-3}$) and corrosion loss of carbon steel ($g\,m^{-2}$).

The selection of appropriate corrosion protection method comprises several steps respecting characteristics of a product, its desired service life and other demands related to its use, corrosive environment and other factors outside the corrosion system, e.g. costs. Relations are shown in Fig. 11.

Corrosion system encompasses both the structural metallic element and its environment – the atmosphere in contact with it. The term atmosphere includes corrosive atmospheric components (gases, aerosols, particles).

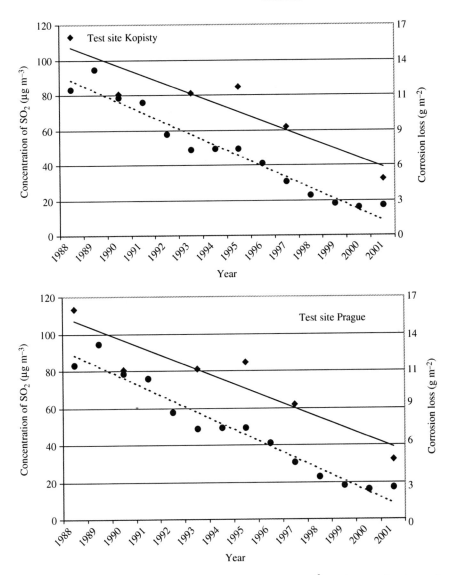

Figure 9: Trend of decrease of SO_2 concentration ($\mu g\,m^{-3}$) and corrosion loss of zinc ($g\,m^{-2}$).

Designed service life is the principal factor in the process of selecting the method of protection for a structural element. The service life of a component or a product is derived in relation to its most important functional property (thickness of an element, noncorroded surfaces, color or gloss). If service life cannot be attained because of shorter life of the selected optimum protection method, it is necessary to carry out one or several maintenance cycles.

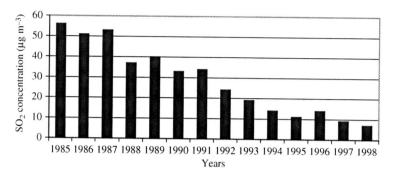

Figure 10: Annual average concentration of SO$_2$ (μg m^{-3}) in the air in Germany in period 1985–1998 [18].

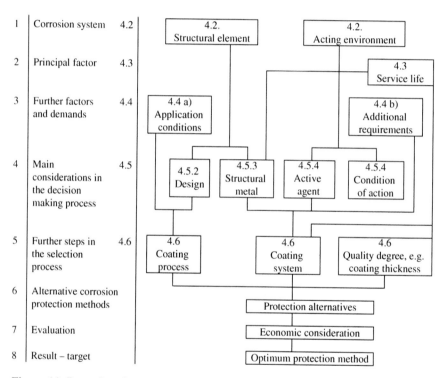

Figure 11: Procedure for selecting a method of protection against corrosion according to ISO 11303.

Further factors to be considered when selecting the protection method relate to technical feasibility of applying the protection and additional requirements derived from the use of a structural element to be protected, e.g. color shade, mechanical or electrical properties, light reflection.

With respect to the component to be protected the main considerations are:

- design;
- structural metals.

With respect to the environment the main considerations are:

- active agent (e.g. gaseous pollutants and particles);
- conditions of action (e.g. humidity, temperature, level and changes, etc.).

Shape, size and other design factors of a structural element exert an important influence on the selection of the optimum protection method. This cannot be described in a generalized form. The influence of design must always be considered individually. Within the corrosion system, the design of a structural element determines the severity of atmospheric effects on individual surfaces, e.g. by different time of wetness, exposure categories or accumulation of corrodants.

Many factors influence corrosivity of atmosphere. ISO 9223 provides means for classifying corrosivity of atmospheres based on four standard metals (carbon steel, zinc, copper and aluminum). The controlling factors are time of wetness and deposition of chlorides and sulfur dioxide (airborne corrodants). The system of ISO 9223 standard does not differentiate conditions in indoor environments of low corrosivity sufficiently since it is not possible in the full extent of classifications of corrosive environments covered. More detailed differentiation and classification of low-corrosivity indoor atmospheres are a part of ISO 11844-1 standard.

Atmospheric environment and exposed metal, or other materials or protective coatings, form a corrosion system to which an appropriate corrosivity category can be assigned. Therefore, various materials are exposed to various stress in a given environment. Surface conditions of a basis metal, e.g. presence of corrosion products, salts and surface roughness, are of a decisive influence on corrosion attack and the durability of protection against corrosion.

The need for protective measures is based on the application of corrosivity categories. For deriving corrosivity categories the basis are either annual corrosion losses of standardized specimens of four basic structural metals or annual means of the most important environmental characteristics affecting atmospheric corrosion. Measured values are ranked into different classification categories and generalize certain ranges of environmental effects on metallic materials (Table 10).

Provided that similar corrosion mechanisms apply, corrosivity categories yield useful information about corrosion behavior of metals, alloys and metallic coatings. Corrosivity categories are not applicable to stainless steels for which data have to be derived directly by taking the main factors of environment and the specific behavior of these steels into account.

Information on corrosivity categories derived from standard metal specimens can also be used for the selection of protective coatings as described in detail in ISO 12944-2 and ISO 12944-5.

The design of a structural element causes smaller or greater variation from standardized surfaces used for specification of corrosivity categories. The orientation of a surface affects corrosion loss.

Table 10: Examples of metallic coatings for corrosion protection of steel structures in atmospheres of different corrosivity.

Typical period before first maintenance (years)	General description	Mean coating thickness (min.) in μm for corrosivity categories			
		C2	C3	C4	C5
Very long (20 or more)	Hot dip galvanized	25–85	45–85	85	115–200
	Unsealed sprayed aluminum	100			150–200
	Unsealed sprayed zinc	50	100	100	150–200
	Sealed sprayed aluminum	50	100	100	150
	Sealed sprayed zinc	50	100	100	150
	Hot dip galvanized sheet	20			
	Zinc electroplated steel	20			
Long (10–20)	Hot dip galvanized	25–85	45–85	85	85–150
	Unsealed sprayed aluminum	100			150–250
	Unsealed sprayed zinc	50	100	100	150–250
	Sealed sprayed aluminum	50	100	100	100–150
	Sealed sprayed zinc	50	100	100	100–150
	Hot dip galvanized sheet	20			
	Zinc electroplated steel	20			
Medium (5–10)	Hot dip galvanized	25–85	25	25–70	45–85
	Unsealed sprayed aluminum	100			100–150
	Unsealed sprayed zinc	50			100–150
	Sealed sprayed aluminum	50			100
	Sealed sprayed zinc	50			100
	Hot dip galvanized sheet	20	25	20	
	Zinc electroplated steel	20	20	25	
Short (under 5)	Hot dip galvanized	25	25	24–45	25–55
	Unsealed sprayed aluminum				
	Unsealed sprayed zinc				
	Sealed sprayed aluminum				
	Sealed sprayed zinc				
	Hot dip galvanized sheet	20	20	20	20–25
	Zinc electroplated steel	20	20	25	25

In general, selection of a coating process may be limited by the design of a structural element (e.g. accessibility may limit use of a spraying process and size may limit the use of a hot dip galvanizing process).

9 Mapping of regional layout of corrosivity of atmospheres

Mapping of regional layout of corrosivity is one of several methods to provide complex but generalized information on corrosion stress on materials and contribute to rational use of materials and selection of measures against corrosion.

The basis for reasonable mapping of corrosivity layout is the knowledge of deterioration processes, meaning kinetics of the process in dependence on occurrence of external factors. Every mapping more or less generalizes the data, which is contrary to the fact that real corrosivity is derived from local environmental conditions and other characteristics of an exposed object forming the 'corrosion system'.

Efforts to map regional corrosivity layout have been going on for several decades [19, 20]. Both methods of deriving corrosivity category have been used during the preparation of the basic data for mapping. Results of site tests of materials (most frequently zinc and carbon steel) were used as well as the information on levels of the main corrosive factors occurring in given regions (sulfur dioxide or NaCl in maritime areas). Both methods were combined in some maps [20]. Maps elaborated for smaller regions provide more realistic data and their results can be used better; however, maps elaborated for big regions are important within more general environmental scope.

The new stage of mapping corrosion effects started in the last decade, especially when activities fulfilling the Convention on Long-Range Transboundary Air Pollution allowed for environmental engineering to include the range of activities aimed at derivation of corrosivity and so cover expression of environmental effects on materials and objects.

The methodology for mapping corrosion effects is included in the mapping manual [21]. Approaches of individual countries to the creation and use of maps of corrosivity layout were presented at the workshop focused on this topic [22].

Mainly places exceeding critical levels are mapped for components of environment. Mapping direct values of decisive components of pollution and corrosion rate values reached in real environments seems to be of technical importance for materials, allowing for a link-up with many provisions of technical standards in the field of protection against corrosion.

Atmospheric degradation of materials is a cumulative and an irreversible process in nonpolluted atmospheres. Therefore, critical stress and critical levels of pollution cannot be defined for materials as for other components of ecosystems. Materials are much more sensitive to deterioration caused by pollution and degradation is accelerated by even minimal occurrence of some pollution components.

Based on the results of a long-term testing program, UN ECE ICP *Effects on Materials Including Historic and Cultural Monuments,* a concept of *acceptable levels of pollution* was elaborated. Acceptable level of pollution is defined as concentration or deposition of a pollution component which does not cause unacceptable acceleration of corrosion or degradation. *Acceptable corrosion rate* or degradation can be defined as deterioration which is technically and economically tolerable. In practice it is expressed as an average corrosion rate multiple in areas reaching 'background' (i.e. minimal) pollution levels (Table 11).

Acceptable corrosion rates K_a are defined as multiples of corrosion rate at background pollution levels K_b:

$$K_a = nK_b,$$

where values of n are 1.5, 2.0 and 2.5. National conditions of actual pollution levels and real possibility for its reduction are considered when selecting n values.

Table 11: Annual corrosion rates of metals in environments at background pollution levels.

Material	Corrosion loss (K_b)	
	$g\,m^{-2}\,yr^{-1}$	$\mu m\,yr^{-1}$
Steel	72.0	9.16
Zinc	3.3	0.46
Copper	3.0	0.33
Bronze	2.1	0.23
Aluminum	0.09	0.03

Elaboration of maps based on direct corrosion tests is organizationally and economically demanding and can be used rather on small regional scales.

Therefore, the main focus nowadays is on using information on levels of important components of atmospheric environments available and transforming these data into values of expected corrosion effects using deterioration equations (dose–response functions) derived from tests at extensive testing site networks. This approach allows for the use of environmental information processed in a comparable manner on an international scale and further processing of these data using GIS technologies.

Use of available information provides elaboration of maps on several levels, which then differentiates the manner of their use. Final maps of occurrence of acceptable corrosion rates and degrees of exceeding these rates are of a fundamental importance to various forms of environmental and economic decisions. Maps of occurrence of decisive factors and calculated corrosion rates allow for estimation or derivation of corrosivity categories and so for appropriate selection of measures against corrosion and estimations of service life. An example for mapping of zinc corrosion rates in the Czech Republic for 2001 is presented in Fig. 12 [23]. Although the information presented by mapping is very valuable [24], it need not be overvalued. Each atmospheric environment has specific local characteristics; therefore, information not provided by maps needs to be used in some cases.

Use of information on regional layout of corrosivity of atmospheres for economic evaluation of damages within an observed area requires the determination of types and quantity of exposed materials [25, 26].

10 Conclusion

Various classification systems have been elaborated during the past decades considering the great importance of data on corrosivity of atmospheres and the opportunity of wide technical use of the data. Newly introduced are a proposal of more sensitive complementing procedures of derivation for the classification system of

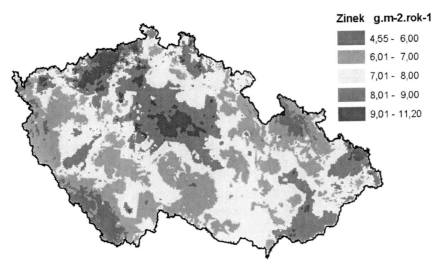

Zinek g.m-2.rok-1

- 4,55 - 6,00
- 6,01 - 7,00
- 7,01 - 8,00
- 8,01 - 9,00
- 9,01 - 11,20

Figure 12: Map of the Czech Republic with the yearly corrosion rates of zinc in 2001 $(g\,m^{-2}\,a^{-1})$.

low-corrosivity indoor atmospheres and use of deterioration equations for derivation of corrosivity from environmental data.

Attention is being paid to assessing probability of occurrence of corrosion in soil and water systems on European scale. New EN standards [27, 28] are rather guidelines for the decision-making system for the evaluation of corrosion occurrence probability in these environments. EN standards for evaluation of corrosion likelihood in the soil of buried metallic structures define the general concepts or the assessment methods and list the main factors influencing the corrosion of buried structures. For new and existing structures from low alloyed and nonalloyed ferrous materials more detailed criteria for soil, backfill materials, environment and structure are given.

References

[1]   CSN 03 8203/ST SEV 991-78 Classification of atmospheric corrosivity.

[2]   EN 12 500:2000 Protection of metallic materials against corrosion – Corrosion likelihood in atmospheric environment – Classification, determination and estimation of corrosivity of atmospheric environments.

[3]   Knotkova, D., Boschek, P. and Kreislova, K., Results of ISO CORRAG Program: processing of one-year data in respect to corrosivity classification. *Atmospheric Corrosion, ASTM STP 1239*, eds. W.W. Kirk & H.H. Lawson, American Society for Testing and Materials: West Conshohocken, PA, 1995.

[4]   Boschek, P. & Knotkova, D., Program ISOCORRAG – repeated one-year exposures. Optimal regression function for adjustment of corrosivity categories. *Proceedings of the UN/ECE Workshop on Quantification of Effects of Air Pollutants on Materials*, Berlin, May 24–27, 1998.

[5]   Morcillo, M., Atmospheric corrosion in Ibero-America: The MICAT Project. *Atmospheric Corrosion, ASTM STP 1239*, eds. W.W. Kirk & Herbert H. Lawson, American Society for Testing and Materials: Philadelphia, PA, 1995.

[6]   Johansson, E., Corrosivity measurements in indoor atmospheric environments: a field study, Licentiate Thesis, Royal Institute of Technology, Stockholm, Sweden, 1998.

[7]   Jones, M.S. & Thomson, G.E., The use of stone thin sections in the study of decay of individual components of stone by pollutants. *Proceedings of the 8th International Congress on Deterioration and Conservation of Stone*, Berlin, pp. 771–782, 1996.

[8]   Bruggerhof, S., Laidig, G. & Schneider, J., Environmental monitoring with natural stone sensors. *Proceedings of the 8th International Congress on Deterioration and Conservation of Stone*, Berlin, p. 861, 1996.

[9]   Vlckova, J., Wasserbauer, R., Chotebor, P. & Kopriva, J., Documentation of sandstone damage on selected parts of St. Vitus' Cathedral in Prague, SVUOM, research report 2/95, October 1995.

[10] Knotkova, D., Vlckova, J., Slizkova Z., Kohoutek, Z. & Wasserbauer, R., Monitoring of environmental effects on built cultural heritage, Envirokor Ltd., research report, 2001.

[11] Romich, H., Leisner, J. & Bohm, T., Monitoring of environmental effects with glass sensors, EC research workshop on *Effects of Environment on Indoor Cultural Property*, Wurzburg, Germany, December 1995.

[12] VDI 3955 Assessment of effects on materials due to corrosive ambient conditions. Part 1: Exposure of steel sheets, Part 2: Exposure of glass sensors.

[13] Tidblad, J., Kucera, V., Mikhailov, A.A., Henriksen, J., Kreislova, K., Yates, T., Stockle, B. & Schreiner, M., Final dose–response functions and trend analysis from the UN ECE project on the effects on acid deposition. *Proceedings of the 14th International Corrosion Congress*, Cape Town, South Africa, 1999.

[14] Tidblad, J., Kucera, V., Mikhailov, A.A. & Knotkova, D., Improvement of the ISO classification system based on dose–response functions describing the corrosivity of outdoor atmospheres. *Outdoor Atmospheric Corrosion, ASTM STP 1421*, ed. H.E. Townsend, American Society for Testing and Materials International: West Conshohocken, PA, 2002.

[15] Morcillo, M., Almeida, E.M., Rosales, B.M., et al., Functiones de Dano (Dosis/Respuesta) de la Corrosion Atmospherica en Iberoamerica, *Corrosion y Proteeeccion de Metales en las Atmosferas de Iberoamerica*, Programma CYTED: Madrid, Spain, pp. 629–660, 1998.

[16] Knotkova, D., Kreislova, K., Kvapil, J. & Holubova, G., Results from the multipollutant programme: Trend of corrosivity based on corrosion rates (1987–2001), draft report of UN ECE ICP Effects on Materials, April 2002.

[17] Knotkova, D., Kreislova, K., Kvapil, J. & Holubova, G., Corrosion rates of carbon steel and zinc in contemporary types of atmospheres. *Proceeding from Conference AKI*, Prague, 2002.

[18] Gauger, T. & Anshelm, F., Air pollutants in Germany: long-term trends in deposition and air concentration. *Proceedings of a UN ECE Workshop on Mapping Air Pollution Effects on Materials Including Stock at Risk*, Stockholm, p. 11, June 2000.

[19] Rychtera, M., Atmospheric deterioration of technological materials, *A Technoclimatic Atlas*, Academia Prague, 1985.

[20] Knotkova, D. & Vlckova, J., Regional appreciation of atmospheric corrosivity on the territory of the Czechoslovak Republic, Guideline, SVUOM, report, 1981.

[21] UN ECE Manual on Methodologies and Criteria for Mapping Critical Levels and Loads and Geographical Areas, where they are exceeded, Chapter IV. Mapping of effects on materials, 2004.

[22] *Proceedings of a UN ECE Workshop on Mapping Air Pollution Effects on Materials Including Stock at Risk*, Stockholm, June 2000.

[23] Knotkova, D., Skorepova, I., Kreislova, K., & Fiala J., Modelling air pollution effects at regional scale. *Proceedings of the International Workshop on Cultural Heritage in the City of Tomorrow*, London, p. 189, June 2004.

[24] Fitz, S., *et al.*, Analysis of risks based on mapping of cultural heritage at different pollution scenario, Study performed within the project SSPI-CT-2004-501609, 2004–2006.

[25] Knotkova, D., Holler, P., Smejkalova, D., Vlckova, J. & Hollerova, P., Evaluation of economic losses caused by atmospheric pollution on historic and modern objects in the territory of Prague, 1991.

[26] Knotkova, D., Sochor, V. & Kreislova, K., Model assessment of stock at risk for typical districts of the city of Prague and Ostrava. *Proceedings of a UN ECE Workshop on Mapping Air Pollution Effects on Materials Including Stock at Risk*, Stockholm, p. 105, June 2000.

[27] EN 12501-1:2002 Protection of metallic materials against corrosion – Corrosion likelihood in soil – Part 1: General.

[28] EN 12502-1:2002 Protection of metallic materials against corrosion – Corrosion likelihood in water conveying systems – Part 1: General.

CHAPTER 4

Atmospheric corrosion and conservation of copper and bronze

D. Knotkova & K. Kreislova
SVUOM Ltd., Prague, Czech Republic.

1 Introduction

Copper and its alloys are important technical metals both for use as construction materials and for special use, e.g. in electro-technical industry and crafts.

This chapter deals with the atmospheric corrosion of copper and bronze used for architectural purposes, constructions and objects of art, especially sculptures [1]. Such uses rely on a number of properties of copper and copper alloys such as high resistance to atmospheric corrosion or ability to form characteristic layers of corrosion products (patina) which not only protect objects, but also usually make objects look better. High resistance of copper and bronze corrosion product layers to corrosion allows for greater stability of objects in the atmospheric environment. Objects and elements made of these materials survive for centuries.

Copper and bronze objects or elements of construction (sculptures, roofs, cladding, artifacts of stone-made sculptural and architectural historical monuments) are affected by atmospheric environment. Its effect, especially with long-term exposure, causes gradual changes. The surface of objects is covered with corrosion products, dirt, biological contamination, etc.

The negative effect is seen mainly in outdoor atmospheric environments where along with natural climatic factors (changes of temperature, relative humidity, precipitation, snow, etc.), gaseous and solid pollutants released by industrial activities are also present. Aggressive components of pollution (gaseous, acid rain) are transported to long distances, which affect materials in a wider area.

Indoor atmospheres (museums, galleries, depositories, etc.) are less aggressive, although in closed spaces in depositories or storage unfavorable corrosion conditions may be formed which can cause damage to the stored copper and bronze objects. This chapter does not specifically deal with this area.

Copper and its alloys also show specific corrosion manifestations in the form of inter-crystalline corrosion and corrosion cracking. Corrosion needs to be eliminated in bimetallic corrosion cells.

This chapter also focuses on studies of defects and disorders in copper and bronze objects exposed to atmospheric environment. Selected technological procedures of cleaning and conserving elements of construction objects namely, historical monuments are presented too.

2 Atmospheric corrosion of copper and its alloys

Copper

Copper is a suitable material for many constructions and architectural elements such as wall cladding, roof and copula roofing. Copper has good mechanical and metalwork properties and is therefore used for construction accessories such as gutters, window ledges, edging, troughs and chimney facing.

Properties of copper are significantly influenced by production technology. For centuries copper, like other metals, has been produced on a small scale by artisans. Metallurgically refined copper contains usually around 99.7% Cu, and silver or other precious metals are not taken out of it. It can be used directly for rolling. Electrolytic refinement produces purer copper containing 99.92%–99.97% Cu, 0.004%–0.02% O_2 and only 0.01%–0.015% of all metal admixtures, the rest being solved gases.

Sheets of rolled copper are used mainly for architectural and construction elements in atmospheric environment. Atmospheric exposure of copper sheets causes change of mechanical properties (strength, ductility) as corrosion spreads over the whole surface exposed to the corrosive environment. After 20 years of such exposure, tensile strength decreases by less than 5% and the change in length is 10% of the original values. Fatigue limit depends on the refinement of crystallization and can go down by one-third or more if crystallization gets significantly coarse.

Copper alloys

Historical monuments and objects of art are made mainly of bronze; brass is used only rarely. Bronze of various compositions is used. The differences among portions of individual alloy elements are big; these also depend on historical trends.

Traditional types of bronze used for making sculptures contain tin, lead and zinc, the share of each of these being approximately 5%. Modern types of bronze are represented by a wider variety of alloys containing approximately 90% copper and other metals such as tin, zinc, lead, aluminum, silver and others. Zinc and tin increase the breaking strength and hardness of an alloy; it improves castability and is cheaper than tin. Lead increases resistance to corrosion and improves the ease of processing and castability. Bronze used for sculptures also contains antimony and a small amount of iron and nickel (tenths of %). When an alloy is cooled slowly, it passes through different phases, each being different in crystalline composition and content of individual elements. The resistance of the types of bronze used nowadays to corrosion does not significantly vary from the resistance of copper, evaluated in terms of corrosion loss. Surface manifestations of corrosion attack are more uneven: dips and corrosion on the grain edges are present.

Currently, bronze with a high level of resistance to corrosion is being developed to be used for making copies of valuable sculptures that should be stored in depositories. The increased resistance of newly developed alloys shall be achieved by additions of tin, silicon, or nickel.

A less-used group of modern bronze alloys is formed by alloys containing tin (10–14%). These materials are close to the historical types of bronze. Their resistance to corrosion is very good, and they are easy to process containing up to 10% of tin.

2.1 Corrosivity of atmospheres and guiding corrosion values for copper and copper alloys

Copper and bronze objects are exposed to atmospheric environment over a long term. It is important to know the conditions of the environment to estimate and evaluate the damage being caused and to select proper protective measures, that is, the means of conservation.

Corrosivity of atmospheres has been rising since the end of the 19th century together with atmospheric pollution, reaching maximum levels in the 1950s and 1960s. There has been a significant decrease during the last ten years as a consequence of various measures against atmospheric pollution and the change of industrial structure.

Standard ISO 9224 – Corrosion of metals and alloys – Corrosivity of atmospheres – Guiding values for the corrosivity categories provide guiding corrosion rates for copper for corrosivity categories according to ISO 9223 – Corrosion of metals and alloys – Corrosivity of atmospheres – Classification (see Table 1).

Corrosion of copper and bronze within Europe was evaluated as a part of the UN ECE International Co-operative Programme on *Effects on Materials including Historic and Cultural Monuments* for the period of eight years (1987–1995) [2].

Table 1: Guiding corrosion rates for copper.

Corrosivity category	Corrosion rate (μm/a)	
	First 10 years	Following years
C1	$\leq$0.01	$\leq$0.01
C2	0.01 < 0.1	0.01 < 0.1
C3	0.1 < 15	0.1 < 1.0
C4	1.5 < 3.0	1.0 < 3.0
C5	3.0 < 5.0	3.0 < 5.0

Table 2: Average mass loss (g/m^2) of copper and bronze on 20 European test sites [2].

Material	Exposure period	Time of exposure (years)		
		1	2	4
Copper	1987–95	10.1	16.4	27.8
	1997–2001	7.6	12.1	19.7
Bronze	1987–95	7.3	12.8	23.6
	1997–2001	4.2	7	12.8

Yearly corrosion loss of copper was between 0.2 μm/year (test site Toledo, Spain) and 1.6 μm/year (test site Kopisty, Czech Republic). Long-term exposure at the atmospheric station of SVUOM (1981–96) in the environment of C4 corrosivity category resulted in a corrosion rate of 2.6–2.8 μm/year. Comparison of corrosion rates of copper and bronze exposed repeatedly in Europe in the first and second stages of the UN ECE International Co-operative Programme shows how the corrosivity of the observed territory has been changing during this time period (Table 2).

Decay of copper and bronze objects is highly influenced by environmental conditions, mainly by acidifying airborne pollutants. In the past ten years environmental conditions have changed in many European countries, the effect being a significant decrease of sulfur dioxide (−75%) and nitrogen dioxide (−40%) concentrations, whereas the amount of ozone remains nearly the same. Sulfur dioxide is still the main factor for corrosion processes but the presence and combinations of other pollutants are becoming more important. Repeated long-term exposure of copper (99% Cu) and bronze (81–85% Cu, 6–8% Sn, 3–5% Zn, 5–7% Pb) samples for 1, 2 and 4 years were carried out to prove and compare the corrosion effect under actual environmental conditions [2].

Compared to the exposure carried out between 1987 and 1991, reduced sulfur dioxide content in the environment seems to influence bronze surfaces more than copper surfaces. The mass loss of bronze reaches values lower by 45%, but copper's metal loss lower only by 24%. The most significant decrease of values can be seen in the weight change of bronze, which reaches values lower by 77% than ten years ago.

The results of the UN ECE International Co-operative Programme were used for the derivation of damage equations for copper and bronze, which help to calculate corrosion loss from environmental data. Besides the usual components of atmospheric pollution, the concentration of tropospheric ozone is considered for the calculation of the corrosion loss of copper and bronze [3].

Long-term corrosion tests of copper and its alloys and the following derivation of guiding corrosion rates are based mainly on corrosion tests using metal sheet samples. Casts made of copper alloys have different properties, which result in differences in the effect corrosion has on them.

Information about corrosion of copper in specific conditions was published in a study [4]. Copper is influenced even by a clear atmosphere at room temperature. A thin layer of invisible oxide is formed; no darkening appears. This layer provides some protection against low levels of pollution in indoor atmospheres.

If copper is exposed to air at high temperatures, a thin coating of interference colors is formed. If traces of sulfides or hydrogen sulfide are present in the atmosphere, copper darkens. Mixed layers of oxides and sulfides are very thin and adherent.

The initial corrosion effect in outdoor atmospheres is staining. It is typical for roofing and cladding and is not considered aesthetically valuable. Sometimes sharply demarcated stains and stripes are formed, especially at places where water is held up and dries out. Stains are dark, dark red or turning blue. This effect is accelerated even by the presence of mild hygroscopic salt deposits.

Stress corrosion cracking (SCC) is typical especially for brass, but is less for other copper alloys. The effect appears under tension stress above a certain limit in a polluted (industrial) atmospheric environment, especially when it contains ammonium compounds. SCC affects copper alloys containing 20% or more zinc, but only rarely the other alloys. Among the objects presented in this publication, brass screws connecting construction elements as window frames, metalwork details, etc. are affected by SCC. In sheltered exposures SCC can be accelerated by cumulating dust and other dirt.

2.2 Formation mechanism of surface layers

Atmospheric corrosion of copper and bronze results from the presence of air humidity and oxygen, while aggressive gaseous and solid components of atmospheric pollution are effective too [5–7]. The process, mainly in its initial stage, is of electro-chemical nature. Since layers of surface electrolytes are thin and dry out periodically, the reaction products are deposited in the form of solid surface layers which then take part in corrosion reactions, and the nature of the corrosion process becomes chemical too. Stability or solubility of surface layers components, especially the ability to bond aggressive atmospheric components into salts of limited solubility, have a fundamental influence on keeping corrosion rates of copper and its alloys low in atmospheric conditions. High resistance of copper and its alloys to corrosion in atmospheric conditions is then derived from the protective function of corrosion product layer usually called patina. Surface layers on bronze and copper are formed by periodically repeated processes during wetting and drop-out of basic salts from saturated electrolytes at a suitable but not too low pH.

Formation of patina is a long-term process going through a number of stages. Different mechanisms determine the speed of the process in individual stages. The corrosion product layer is subject to transformation in the formation period and eventually it gets into balance with the external environment affecting it. The natural, stable, green surface layer is formed in 10 to 20 years in atmospheric conditions, depending on the corrosivity of the atmosphere. At low levels of pollution in unsuitable climatic conditions (dry, low temperatures) the green surface layer is formed in a long time or could not be formed at all. Surfaces remain dark. Different local

Table 3: Stability of basic patina components in pH levels [7].

Compound	Formula	pH level
Cuprite	Cu_2O	>4
Brochantite	$Cu_4(SO_4)(OH)_6$	3.5–6.5
Antlerite	$Cu_3SO_4(OH)_4$	2.8–3.5
Atacamite	$Cu_2Cl(OH)_3$	3.8–4.3
Gerhardite	$Cu_2(NO_3)(OH)_3$	4.0–4.5
Malachite	$Cu_2(CO_2)(OH)_2$	>3.3

degrees of wetness and other microclimatic effects cause uneven coloring of objects, e.g. roofing and sculptures. Typical properties of surface layers can be estimated after 3–5 years of exposure.

Primarily, a layer of cuprite is formed on newly exposed surfaces of copper and bronze after wetting. Surface electrolytes in the next corrosion stage contain sulfate ions corresponding with their dry and wet depositions from the atmosphere and ions of copper diffusing through the cuprite layer. Copper sulfates turn into basic sulfates, which are deposited on the cuprite surface in the form of solid layers.

Formation and composition of patina are significantly influenced by acidity of rain. Rain of high acidity dissolves the surface layer and washes it off. Rain of usual acidity influences the composition of patina in its pH, because the thermo-dynamical stability of minerals differs, e.g. antlerite is more stable in acid environments than brochantite (see Table 3).

Basic patina components are formed in the presence of surface water layer by the nucleation and deposition of solid phases in a row of reactions. Brochantite is said to need at least two weeks to crystallize. During this period insufficiently bonded copper and sulfate ions can be washed out by rain. Wash-out is effective especially for thin initial layers.

Relations between time required for the deposition of more stable components of patina and the occurrence of rain may influence the formation of layers of natural patina.

2.3 Influence of selected factors on quality of forming layers

Patina on copper or bronze object differs locally in thickness and composition in visually distinctive areas. The reason for this variety is the shape arrangement of objects, which creates differences in the access of individual environmental factors to the material in various parts either as a consequence of un/shelteredness from the factors or various speeds at which environmental components are transported to the surface. The surface layer formed manifests as various shades of green, can be blue and green, gray and green, brown or black. Results of X-ray diffraction analysis show that color (especially dark) does not provide sufficient information on composition of a layer. Color borders can be sharp or diffusive.

Occurrence of green stripes formed by streams of water flowing away on dark background is typical for sculpture surfaces. Dark, coarse and thick crusts on partial surfaces, mostly on soffits, usually differ in appearance too. These crusts contain many foreign matters – impurities and crystallized soluble fractions of corrosion products of copper. The protective function of such layers is limited.

Corrosion attack on bells 100–600 years old that was exposed in various atmospheric environments in the Czech Republic (countryside, cities, highly polluted industrial areas) was evaluated [1]. All bells were made of bronze containing a relatively high portion of tin (approx. 20%). Considering the massiveness of the bells, corrosion damage does not represent any threat to the function of the objects. Surfaces are covered with a relatively coherent but not compact layer of corrosion products. Corrosion attack on sides facing wind is usually more significant. No corrosion pits occur. Different manifestations of corrosion can be seen on the inner surfaces of the bells, especially in the bells exposed for long periods. The corrosion layer contains less impurities, is usually blue-and-green, and crusts are present locally. Condensation and wet corrosion probably take place inside the bells. Surface layers contain mainly antlerite, and a high portion of quartz, gypsum and calcite (dust). Also present is moolooite (copper oxalate) which proves biological pollution (birds). Brochantite was also found on a bell exposed since 1651.

Properties of layers, which then affect the corrosion rate of copper and its alloys in the long term, are formed according to conditions of pollution and temperature-humidity characteristics of an area, but also depend on the exposed surface.

The natural surface layer characteristic of current copper and bronze surface exposed for tens or hundreds of years in an open atmosphere contains mainly brochantite, antlerite and cuprite. Formation of antlerite or brochantite is subject to free access of precipitation. In atmospheres with natural or technological salinity, patina also contains atacamite and paratacamite – $Cu_2Cl(OH)_3$. Patinas that have not reached a stabilized state contain posnjakite – $CuSO_4 \cdot Cu(OH)_6 \cdot H_2O$. Gerhardite – $Cu_2(NO_3)(OH)_3$ is found rarely. Patinas nowadays do not contain malachite – $Cu_2(CO_3)(OH)_2$ which was present in analyses from the past years. Surface layers often contain admixtures of gypsum and siliceous sand, and patinas also contain portions of formates, acetates and oxalates which are formed in atmospheres polluted by anthropogenous and biological effects. These components have been identified only in the last few years by the use of ion chromatography in this field [8]. Salts of organic acids are reported to form a kind of binding agent in patina layers. Other substances can be formed in environments with specific effects. Soluble $CuSO_4 \cdot 5H_2O$ (chalcantite) cumulates in soffit areas. An overview of the occurrence of individual components and their properties is presented below.

Some papers refer to other types of components of patina in the initial stage of corrosion too.

2.4 Identification of individual corrosion products

An overview of the phases determined repeatedly in surface corrosion layers on copper and bronze objects in the Czech Republic is presented in Tables 4 and 5.

Table 4: Phases identified in patina in the atmospheric environment of the Czech Republic.

Name	Formula	Line
Antlerite	$CuSO_4 \cdot 2Cu(OH)_2$	7-407
Brochantite	$CuSO_4 \cdot Cu(OH)_2$	13-398
Posnjakite	$CuSO_4 \cdot Cu(OH)_6 \cdot H_2O$	20-364
Cuprite	Cu_2O	5-667
Tenorite	CuO	
Atacamite	$Cu_2(OH)_3Cl$	25-269
Paratacamite	$Cu_2(OH)_3Cl$	25-1427
Botallackite	$Cu_2(OH)_3Cl$	8-88
Moolooite	$C_2CuO_4 \cdot nH_2O$	29-297
	$(NH_4)_2Cu(SO_4)_2 \cdot 6H_2O$	11-660
Chalcantite	$CuSO_4 \cdot 5H_2O$	11-646
Gerhardite	$Cu_2(OH)_3NO_3$	14-687
Melanothallite	Cu_2OCl_2	35-679
	$\beta\text{-}Cu_2S$	26-1169
	Cu_2S	23-959
Quartz	SiO_2	33-1161
Gypsum	$CaSO_4 \cdot 2H_2O$	33-311
Calcite	$CaCO_3$	5-586

Individual components were determined by a phase analysis of samples taken from objects in Prague and other places within the Czech Republic (approx. 300 samples) [9–10]. Components shown in Tables 4 and 5 are typical for urban and industrial atmospheres without any significant salinity effect.

2.5 Composition of surface layers and patina components characteristics

- Brochantite $CuSO_4 \cdot Cu(OH)_2$, antlerite $CuSO_4 \cdot 2Cu(OH)_2$ – corrosion products of copper and bronze are mainly brochantite, antlerite or a mixture of both. Brochantite occurs mainly at places of direct impact of rain, snow and dew, while antlerite is formed mainly by condensed air humidity. Both products have protective properties and slow down further corrosion of metal surfaces. From the aesthetic point of view, they give patina blue-and-green coloring, which, however, can be devaluated by minimum amounts of darker components as carbon black, black oxides and sulfides.
- Cuprite Cu_2O, tenorite CuO – these represent an anchoring component on the metal-patina border. They form thin adhesive layers, mostly black, brown red or orange.
- Posnjakite $CuSO_4 \cdot Cu(OH)_6 \cdot H_2O$ – it is a component of young sulfate patina and turns into antlerite or brochantite later on.

Table 5: Composition of patina layers.

Composition	Objects					
	Statues outside Prague (investigation in 1994) 18 samples		Statues, copper roofs in the center of Prague (investigation in 1994) 57 samples		Statues in the center of Prague (investigation in 1998) 33 samples	
	Abundance	Frequency (%)	Abundance	Frequency (%)	Abundance	Frequency (%)
Antlerite	9	50	18	31	10	30
Brochantite	6	33	43	75	17	52
Cuprite	3	17	16	28	8	24
Atacamite, paratacamite	0	0	3	0	1	3
Moolooite	0	0	0	0	1	3
$(NH_4)_2Cu(SO_4)_2 \cdot 6H_2O$	0	0	4	7	4	12
Chalcantite	0	0	0	0	3	9
Quartz	12	67	33	58	18	55
Gypsum	4	22	2	4	13	40
Calcite	0	0	3	5	4	12
Entirely amorphous	2	11	1	2	0	0

- Atacamite, paratacamite $Cu_2(OH)_3Cl$ – dominant component of patina in maritime environments. Only exceptional occurrence was identified within Czech Republic in cases of presence of chlorides (e.g. surfaces of fountains where chlorides are contained in water or on surfaces affected by de-frosting agents).
- Moolooite $C_2CuO_4 \cdot nH_2O$ – its occurrence is attributed to the effect of bird excrements on metal. It was identified at locations sheltered from a direct impact of precipitation.
- Ammonium copper sulfate $(NH_4)_2Cu(SO_4)_2 \cdot 6H_2O$ – water-soluble corrosion product of copper which is contained in crusts. It was identified at locations affected by anthropogenic influences.
- Chalcantite $CuSO_4 \cdot 5H_2O$ – was identified as a part of soffit crusts in highly polluted environments. This component is well soluble, has no protective effect and its presence indicates increased corrosion of metals.
- Quartz SiO_2 – most frequently occurring component of dust, chemically inert component of surface layers. It increases volume of patina at higher concentrations, decreases adhesion to surfaces and deteriorates the look of patina.
- Calcite $CaCO_3$, gypsum $CaSO_4 \cdot 2H_2O$ – calcite is often a part of dust, having a negative effect on it. It turns into gypsum by the effect of sulfate ions, causes expansion of volume and degradation of patina. High amounts of gypsum were identified in patinas from locations polluted with dust from neighboring lime works.

Besides copper, bronzes contain other alloy elements (lead, zinc, tin, iron). Corrosion products of these components occur in surface layers in small amounts that have no significant effect on the protective properties of these layers. Corrosion products of tin ($SnO_2 \cdot nH_2O$, SnO_2) used in bronze alloys in the past increase protective ability of surface layers.

Surface corrosion layers including the protective patina on copper and bronze do not have a simple composition, not even an unambiguous crystalline structure. Metallographic cuts document two- or three- layer construction of patina in most cases, in some cases foreign matters can be seen in the top green layer. Results [1, 11, 12] are related to the evaluation of layers on copper sheet; surface layers on bronze can be evaluated less frequently [13].

The closest zone to the metal on the cut is orange, red or dark layer, mostly made of copper oxides. The next layer is bright green and contains mainly brochantite, but also antlerite. In this sub-layer, dust and other impurities accumulate. However, siliceous particles can be in the bottom oxide layer too. Dark top layer, if present, is formed mainly by carbon black. Varying and atypical composition occurs in case of voluminous dark or gray and green crusts in soffit locations (high portion of dust particles, antlerite, various soluble corrosion products of copper).

Individual layers, even those that are relatively chemically homogenous, are formed by irregular crystalline formations. Besides impurities, cavities and other defects are also present in layers that allows for the transportation of ions through layers.

The thickness of green patina (green top sub-layer) formed in an open atmosphere in long term is between the range of 40–50 and 75 μm. The thickness of the black, red or dark sub-layer is approximately 10 μm. On the bottom side of the sheets, the dark sub-layer is thicker (approx. 20 μm), while the green sub-layer is only a few micrometers thick. The depth of pits (if they occur) on surfaces exposed in long term ranges from 50 to 70 μm, but also from 100 to 105 μm in some cases.

The thickness of crusts is higher (approx. 200 μm). The boundary of metal surface under these crusts is usually very uneven, broken, with corrosion penetrations of various depths. Penetrations have shapes rather different from those of shallow pits.

To sum up, the surface layer of copper is a system that may vary in chemical composition, composition in the layer cut, protective function and appearance.

3 Typical corrosion manifestations on copper and bronze objects

Copper and bronze objects or construction parts as roofing, cladding, doors or sculptures have been exposed to the atmospheric environment for centuries without any serious corrosion problems in most cases, except for the aesthetic problems caused by cumulating impurities and the formation of corrosion layers differing in color locally. In contrast to copper roofing and cladding, where corrosion damage may affect the service life of an object, damage in bronze sculptures is considered aesthetically desirable both for relatively higher thickness of the exposed material and for periodical maintenance and protection of some sculptures [1, 9, 14–16].

Standardization of defects and disorders and evaluation of causes of their occurrence are listed individually for sculptures (bronze and copper) and for roofing, cladding and metal work after general characterization, the reason being not only partial differences in causes but also differences in technical and aesthetic demands.

3.1 Sculptures and objects of art

Corrosion manifestations (described in more detail below), which may lead to defects in some cases, can be generally divided into the following groups:

- Manifestations of general corrosion differentiated by space arrangement of an object,
- uneven or local corrosion attack which may be resulting from space orientation of a surface but also from construction design (e.g. joints),
- mechanical damage (cracks, changes of shape),
- defects resulting from the way sculptures were made (connection lines, uncovered rivets),
- corrosion of carrying and supporting construction, even bimetallic,
- corrosion damage and formation of deposits resulting from the function of an object (fountains), and
- negative aesthetic manifestations resulting from previous conservation and various surface treatments (e.g. pigment coatings).

Main types of corrosion manifestations:

1. Formation of black and green layers of corrosion products. The composition of most samples of surface layers from various objects corresponds with the typical composition of patina formed nowadays in urban and industrial areas (brochantite, antlerite, cuprite, admixtures of quartz and calcite). The surface layer is bound firmly to the surface. The surface layer is protective, i.e. it decreases the corrosion rate of a metal.

2. Crusts that are formed mainly in sheltered locations are of usual composition (admixtures of gypsum, quartz, etc.). Specific components as ammonium copper sulfate $(NH_4)_2Cu(SO_4)_2 \cdot 6H_2O$ and copper formate $((HCOO)_2Cu)$ were found only in a small number of cases. Crusts on sculptures located in the center of Prague often contain chalcantite $(CuSO_4 \cdot 5H_2O)$ and moolooite, i.e. copper oxalate $(C_2CuO_4 \cdot nH_2O)$. These fractions do not have a protective function.

3. Turquoise areas are formed around little pits on surfaces of bronze sculptures where condensate from inside the sculpture reaches out.

4. Uneven corrosion along grains was observed in a limited number of cases on horizontal and less inclined surfaces of statues exposed in highly polluted environments.

5. An interesting finding was the occurrence of places with little pits in overlapping areas on sculptures made of hammered copper sheets (e.g. under coat overhang on a statue of a horse rider). These pits were under a layer of deposited dirt and corrosion products. It may be assumed that it is a case of crevice corrosion.

6. Bimetallic corrosion of steel joints and fixation elements and their surrounding was found on parts of sculptures that are not fully treated (back side). Corrosion of steel joints and voluminous corrosion products of iron may damage and open the joints.

7. Extensive corrosion damage was found on the inner steel supportive construction of sculptures made of hammered copper sheets. Bimetallic corrosion occurs at places where steel inner construction is in contact with copper surface. Corrosion attack on inner supporting construction reaches the degree of defective stability of an object in some cases, which even increases penetration of environment into the inner space of a sculpture.

8. Various joints and elements used during the treatment of a sculpture become visible (joint lines, matching of parts, overlaps of inner joints).

9. Soldered joints on sculptures made of copper sheets are being damaged and local corrosion manifestations occur around these joints. Dark dots and stains occur on soldered areas. This is also a case of soldered decorative parts of cladding and copulas.

10. Cracks in surfaces and deformations or fall-off of parts are rather rare.

11. Corrosion manifestations related to aging of sculpture coatings (conservation, patination, pigmented and transparent coatings) are aesthetically undesirable.

Coatings change in color, wash away, grow through corrosion products or are gradually separated from the base.

12. Water deposits in fountains.
13. Corrosion manifestations on bells are significantly influenced by the manner of exposure (tower spaces) and shape (inner and outer surface). Pollution by bird excrements is an important factor. Serious damage occurs after mechanical wear-out on the inner surface and corrosion damage on steel parts of bell hangers.

3.2 Copper roofs, cladding and metalwork elements

Copper sheets and metalwork elements were used in the past only for roofing and other parts of important buildings (churches, palaces). During the last ten years there has been a growth in the use of copper for a wider range of purposes (family houses, business objects).

Both the types of use differ in the material used, construction design and technological treatment which is reflected by differences in the tendencies for formation of defects and damages. Material and technological differences of both types of use are also reflected in the mode of maintenance and repairs.

A great deal of attention has to be paid to the designing and making of copper roofs or claddings of a modern object. Designing copper (and also titan-zinc) roofing and cladding is a specialized technical work nowadays. Producers of materials and elements elaborate extensive manuals setting construction designs and technological treatment of elements, their connections and making them into an object, especially in relation to its temperature-wetness regime and material composition in combination with copper parts.

There are four main causes of defects and damages of copper roofs and metalwork elements – mechanical damage, corrosion impact of the outdoor environment, corrosion impact derived from the construction design of a building and specific corrosion impact derived from its indoor regime. The degree of stress caused by individual effects may differ, and causes are often combined. The method of repair is chosen according to the type of defect.

Mechanical damage
- parts loosen by wind,
- extensive heat stress, changes of temperature,
- obstacles to compensate movement during heat stress,
- operation stress (walking areas),
- accidental damage (falling objects, perforations).

Corrosion attack (environmental effect and design)
- impact of surrounding local atmospheric environment,
- local effect of smoke,

Table 6: Types of defects on bronze and copper objects in Prague.

Age of objects (years)	Type of object		Defect		
	Roof, cladding (copper)	Sculpture (bronze, copper)	Aesthetic[a]	Corrosion[b]	Mechanic[c]
450–240	1	5	3	3	2
150–110	–	4	2	1	1
100–80	3	28	25	21	4
75–60	3	17	13	13	6
50–15	1	39	24	24	6
<15	2	4	3	5	1
Total	10	97	70	66	20

[a]Nonuniform patina, differences in color.
[b]Pitting, crusts, etc.
[c]Holes, damaged joints and supporting constructions.

- corrosion effects caused by contact with unsuitable materials (bimetallic corrosion, soldering),
- corrosion effect of rinsing of surrounding materials (acidic fractions and others),
- corrosive and other negative effects of nesting birds, mainly pigeons,
- pitting on roofing overlaps and board welts (depth to $50\,\mu m$),
- formation of non-protective surface layers (ammonium copper sulfate contained) around ventilation inlets where exhalations from residential and operation spaces of buildings leave,
- rusty areas on roofs surrounding steel elements (e.g. lightning-conductor connections).

Damages caused by indoor effects
- condensation,
- indoor exhalations,
- damaged base or inner construction as a result of decay or insect attack.

Table 6 gives an overview of the different types of defects on copper and bronze objects in Prague [1, 14].

4 Selected methods of cleaning and conserving copper and bronze objects

This section lists methodical knowledge, conditions of use and properties of treated surfaces for these technologies:

- physical (mechanical) and chemical cleaning methods,
- patination and properties of artificial patina including their transformation after exposure in atmospheric environment,
- conservation.

The main benefit is quantification of changes caused by selected technologies of cleaning on surfaces of objects [1].

4.1 Cleaning of copper and bronze objects, namely sculptures

Techniques of cleaning and removal of corrosion products [1] can be divided into three groups:

- cleaning by water under pressure,
- mechanical or abrasive cleaning (blasting),
- chemical cleaning – 'draw-off' and pickling.

4.1.1 Cleaning by water under pressure

Removal of corrosion products and dirt by water under pressure is often used in combination with abrasive techniques or chemical cleaning. Pressures between 50 and 1000 psi are used for water cleaning. Particles that fall off or are not fixed to the surface and soluble parts of corrosion crusts and layers are removed or redistributed on the surface by pressure washing. The effectiveness of washing depends mainly on the conditions and configuration of a surface and the abilities and experience of staff.

4.1.2 Mechanic (abrasive) cleaning

Various methods of mechanical cleaning such as different types of abrasive and polishing agents, pastes, metal wool and brushes made of various materials including mechanical brushes, scalpel, special abrasive steel wool, etc. are used by conservators. Cleaning using hand-operated mechanical equipment is difficult and time-consuming. These methods can remove crusts, deposits and growth from a surface and keep a thin layer of patina. Whether it is possible and desirable to maintain a thin layer of patina depends on the degree of pollution of a sculpture, previous surface treatment, shape of an object, etc. Usually not all parts of a sculpture can be cleaned in this way, especially those with no access.

A big problem during the mechanical cleaning of sculptures is the removal of a layer of previous improper 'conservation' by various coatings and waxes or the removal of the remnants of partially deteriorated previous conservation.

Mechanical cleaning using abrasives – blasting – is used on historical objects rather rarely, but there are cases when this method is the one to be used (e.g. inner surface of fountains, bell surfaces). Blasting is a process during which solid particles of metals, minerals, synthetic resins or plants, sometimes combined with water, are thrown at high speed against objects in order to clean, reduce or harden

Table 7: Material loss of copper with a thin brown layer of corrosion products after blasting by various abrasive agents (blasting normal to the surface).

Abrasive agent	Pressure (MPa)	Nozzle diameter (mm)	Material loss			
			Copper		Brass	
			(g/m^2)	(μm)	(g/m^2)	(μm)
Corundum 120	0.4	1	1.5	0.18	1.8	0.22
		2	4.5	0.55	3.5	0.43
	0.2	1	1.1	0.13	0.7	0.09
		2	2.1	0.26	2.9	0.36
Black corundum 180	0.2	1	1.0	0.12	2.2	0.26
Balotine 60	0.2	2	0.5	0.06	0.1	0.02

their surfaces. Compressed air (pressure between 140 and 550 kPa) and abrasives of different sizes, usually between 175 and 1000 μm, are used for blasting. Different combinations of speed, distance from surface and angle of incidence are used. The selection of these parameters is determined by the character of the surface to be cleaned, especially by its coarseness, hardness and thickness of crusts and corrosion layers. These methods have a great impact on the topography of a surface, its deformation and the deformation of microstructure, etc., although very soft abrasives are used, e.g. glass balls, sodium carbonate, aluminum oxide, olivine sand, plastic with different hardness and size of particles, some natural products such as ground walnut shells or crushed corn.

Blasting gives rise to unstable surfaces sensitive to humidity, which makes corrosion processes continue. These methods remove neither the active centers of chlorides or the sulfates from deep surface pits. The blasting process can leave parts of abrasives on a surface, which can negatively affect it after exposure to environment. Aggressive abrasive techniques can uncover more active pits and centers on surfaces than less aggressive procedures. The impact of blasting on surfaces of copper and its alloys was evaluated on samples of copper sheet at a different degree of corrosion attack (Tables 7 and 8).

Surfaces are changed after blasting in a way resulting from the technology used:

- Blasting agents significantly increase the coarseness of cleaned surfaces depending on the hardness of the agent and the shape of its particles.
- Bigger changes of macrostructure are caused by corundum, but the surface is cleaned more efficiently; blasting by balotine leaves remnants of surface layers.
- Blasting contaminates surfaces with elements or impurities contained in the blasting agent. Higher portion of trace amounts of impurities occurs when balotine is used.

Table 8: Weight and thickness loss of copper with natural green patina exposed for 60 years after blasting by various abrasive agents (blasting normal to the surface).

Abrasive agent	Pressure (MPa)	Nozzle diameter (mm)	Material loss	
			(g/m^2)	(μm)
Corundum 120	0.4	2	1.2	0.15
	0.2	1	11.2	1.38
		2	1.0	0.13
Black corundum 180	0.2	1	11.2	1.37
Balotine 60	0.4	2	1.2	0.15
	0.2	2	1.0	0.13
Balotine 134	0.2	1	10.7	1.31

4.1.3 Chemical cleaning

Methods of chemical cleaning include mainly drawing-off and pickling using different chemical solutions. The difference between drawing-off and pickling is that pickling removes all layers of corrosion products down to pure metal.

Drawing-off is a method that can be carried out in standardly equipped workshops or even in the field. It is not quite possible to achieve the defined and even effect of agents on a sculpture surface. Corrosion products are removed only partly and locally which however might be suitable for some purposes.

Chelation or complex-forming solutions or pastes are used for drawing off, e.g. [17]:

- alkaline Rochelle salt (sodium potassium tartrate) removes corrosion products without affecting the base metal;
- applications based on EDTA (ethylenediaminetetraacetic acid) selectively remove unadhesive or active corrosion products from bronze surfaces, leaving them enriched with copper oxides allowing for future repatination;
- application of ammonium hydroxide solution and simultaneous brushing with wool removes unadhesive layers of corrosion products,
- application of a solution of water and ammonium hydroxide (1:1) and simultaneous brushing with a metal brush or pumice removes thick incrustations;
- application of a paste containing 1 volume part of powder soda, 5 volume parts of calcium hydroxide and 2 volume parts of sawdust removes corrosion products for strongly deteriorated and corroded surfaces.

Pickling is a process used in technical practice especially as a pre-treatment before the application of various types of protective coatings. Cleaning is thorough but surfaces are often more affected. Loss of base metal during chemical cleaning processes equals to 2.5–9.0 years of exposure in an urban atmosphere. For technical purposes, pickling is defined in ISO 8407 Corrosion of metals and

Table 9: Mass loss of copper with thin layer of oxides after pickling in various solutions (temperature 20°C).

Pickling solution	Concentration (M)	Time (min)	Total mass loss of copper	
			(g/m^2)	(μm)
Citric acid	0.10	5×3	0.45	0.05
Chelatone 3	0.20	4×20	0.40	0.04
Sodium hexametaphosphate	0.50	4×3	0.28	0.03
Sulfuric acid	0.10	5×2	0.43	0.05
Sulfamidic acid	0.05	5×2	0.85	0.10
Phosphoric acid	5.00	4×2	0.73	0.08
Sodium hydroxide[a]	0.25	4×10	0.48	0.05

[a]Temperature of solution 70°C.

alloys – Removal of corrosion products from corrosion test specimens. The standard includes recommended solutions for individual metals and alloys. According to this standard, pickling periods differ for different pickling solutions, metals or degrees of corrosion. When pickling is used on historical objects it is recommended to apply pickling solution repeatedly, thoroughly wash the objects after each application and remove loosened corrosion products and dirt, mechanically.

Comparison of pickling agents and procedures for copper and brass was carried out both from the point of view of corrosion loss and morphology of surfaces after pickling in solutions (citric acid, Chelatone 3, sodium hexametaphosphate, sulfuric acid, sulfamidic acid, sodium hydroxide, phosphoric acid) [18]. Pickling procedures were verified on copper and brass samples with thin layer of corrosion products formed in indoor environment (Table 9), and copper samples with thick layer of green corrosion products formed after exposure in condensation chamber with SO_2 and salt spray (layer of approx. 0.7 g/m^2 – Table 10) [1]. It is important that during pickling undesirable layers of corrosion products are removed and the base metal is affected to the minimum.

Intermediate loss of copper after individual intervals of pickling by Chelatone 3, sulfuric acid and sodium hydroxide are shown in Fig. 1. Intervals of pickling are defined in Table 9. It is obvious that after pickling in Chelatone 3 solution and sulfuric acid was finished, all corrosion products were removed from the surface (curve gradually turns into straight line). In the case of pickling in sodium hydroxide solution, not all corrosion products were removed after four intervals. All corrosion products were removed by pickling in sulfuric acid already after three intervals (6 min); the base metal has been slightly dissolved in the following intervals. Chelatone 3 solution dissolved corrosion products more slowly, but after four intervals (80 min) all corrosion products were removed.

Table 10: Mass loss of copper with relatively thick layer of corrosion products after pickling in various solutions (temperature 20°C).

Pickling solution	Concentration (M)	Time (min)	Mass loss of copper (g/m²)	(μm)
Citric acid	0.1	5 × 3	22.9	2.6
Chelatone 3	0.2	4 × 20	36.5	4.1
Sodium hexametaphosphate	0.5	4 × 3	56.3	6.3
Sulfuric acid	0.1	5 × 2	4.0	0.4

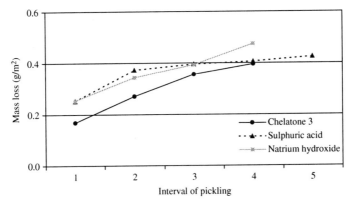

Figure 1: Corrosion loss of copper during pickling.

In the case of the samples covered with a thick layer of corrosion products, the course of pickling is similar: only the absolute values of material loss are higher. Chelatone 3 solution acts more strongly in the first interval, but after the second interval the intensity of dissolving corrosion products is similar to that of the sulfuric acid solution. The total material loss after pickling in Chelatone 3 solution is lower than in the case of pickling in the sulfuric acid solution, while the degree of cleaning the surface is the same (Table 11).

Surfaces are changed after pickling in a way resulting from the technology used:

- etching of surfaces after pickling does not increase the coarseness of surfaces, but the structure of metals is more marked, and
- trace amounts of elements contained in pickling solutions (Na, P) can be found on pickled surfaces even after proper wash-off in some cases.

4.1.4 Other cleaning technologies
Other options are use of high-pressure steam or blasting with little dry ice balls. Electrolytic methods are suitable for cleaning copper and bronze objects, especially

Table 11: Condition of copper surface after pickling in various solutions.

Pickling solution	Condition of copper surface after pickling
Chelatone 3	Original corrosion products removed, products of reaction with the pickling solution present, their occurrence is related to crystallographic orientation of grains of the base metal
Sodium hexametaphosphate	Areal removal, grain borders slightly marked, occurrence of undefinable stains which can be a result of selective pickling or products formed during pickling or material impurities
Sulfuric acid	Original products removed, course of pickling and also formation of new products depends on crystallographic orientation of grains of the base metal
Citric acid	Original products removed, course of pickling and also formation of new products depends on crystallographic orientation of grains of the base metal, less marked than in case of sulfuric acid pickling

archeological findings. The aim of cleaning is to remove aggressive ion fractions (chlorides) from surface layers. These methods allow to keep the layer of patina and other components on a treated object.

4.2 Artificial patination

Artificial patination is an important step in the process of creating and restoring bronze and copper historical objects and objects of art [19]. Artificial patination is not suitable for big outdoor surfaces as copper roofs or copper parts of cladding of buildings. Because of the long time it takes for a natural green patina to form there is a need to create patina artificially. Artificial patination is used mainly when a desired look of a surface needs to be reached quickly. Protective properties or the resistance to corrosion of created layers vary; in most cases they are not known or considered.

4.2.1 Creation, composition and properties of artificial patinas

Natural and artificial patinations are processes differing in principle and so they need to be understood and evaluated. Possibility of use of artificial patina needs to be assessed case by case.

There are a number of formulas for patination containing various components. Most of them are complex procedures, and their success is conditioned by experience and practice. The effect of chemical components is completed by other procedures that mechanically remove unadhesive portions of patina, mechanically increase the homogeneity of remaining layers or cause secondary reactions at raised temperature. The process of creation of artificial patina is fairly difficult. Development of methods of artificial patination is supported by the need for fast formation of relatively evenly green or differently colored surfaces. Requirements for artificial patinas, especially aesthetic ones, are highly specific; treated surfaces are usually not very big and various craft procedures may be used to finish the surfaces. Data on composition of artificial patina are limited, especially data mentioning the patination process together with the final composition of a layer of artificially created products.

Patination is a process of controlled corrosion followed by further treatment of the corroded surfaces. Patination causes loss of the base material; all types of patination cause corrosion loss of bronze several times higher than the average corrosion loss in urban atmosphere (3 to 10 times higher). The highest corrosion loss is caused by two-step procedure (e.g. black patina created by solution of sulfurated potash, green patina created by solution of copper nitrate). However, corrosion loss of 3–10 μm is not important for loss in the thickness of either sculptures or sheets of objects of art on an object.

Mixtures of different chemical substances are used for patination, and creation of aesthetically desired even surface layers of artificial patina is very complicated. Brown and black patina is created in water solutions of sulfurated potash and ammonium sulfide (amorphous mixture of polysulfides). A thin layer of copper sulfide is formed on surfaces of bronze objects. Green patina is usually created in water solutions of copper nitrate or chloride together with other oxidants and complex-forming agents. There are also industrial processes of patination of monuments or semiproducts (sheets).

Longer service life of layers created this way is reached by consequent application of waxes, microcrystalline waxes or transparent coatings. Regular maintenance must be carried out to preserve the final aesthetical effect. However, repeated conservation causes changes in color.

Artificial patinas are more porous and less compact than natural patinas. Layers of artificial patina cannot be sufficiently bound to the copper or bronze base of an object. That is reached gradually by the repeated impact of climatic conditions, especially precipitation and dry-out. Uneven impact of outdoor environment also creates color tones typical for natural patina.

Artificially created green patinas on restored historical objects have a number of disadvantages:

- insufficient adhesion,
- partial wash-out by precipitation (artificial patinas contain more water-soluble components), and
- undesired changes of look during the process of transformation of artificial patina into patina the composition of which is closer to natural patina.

Artificial patinas vary in composition, and the contained components depend on the composition of patination solutions or pastes used and the manner of consequent surface treatment (mechanical treatment – brushing, washing, raised temperature).

4.2.2 Artificial patinas produced in workshops

Several techniques are used for application of artificial patinas [20]:

- Direct patination – direct application of chemical solutions on surfaces of objects using moistened cloths, soft brushes, etc. so that a thin coating is formed on a surface,
- Patination in vapors – objects are exposed to impact of chemical vapors in a closed environment. Areas differing in color can be formed during patination on surfaces of different orientation where gaseous vapors are condensed.
- Dipping patination – the whole object is dipped into chemical solution, often at higher temperatures. Solutions contain oxidizing substances, copper compounds and also acids or alkaline substances in some cases. On big objects, solution can be applied by pouring; however this method does not produce sufficiently homogeneous and even-in-color patina layers.
- Patination at higher temperatures – a wide variety of mainly dark colors of copper and bronze can be produced by the use of heat only. Patina layer is relatively thin. Such dark layer is often used as preparation treatment before further patination to green color.

Surfaces of objects can also be sprayed with suspension of basic copper salts. Suspended particles are then fixed to a surface by wax-binding agents applied after the suspension dries out.

The biggest problems occur in the case of patinating only parts of objects after repairs, exchanges of damaged parts, etc. Even if the color of artificial patina is the same as the color of natural patina, the new layer becomes visible due to the transformation of artificial patina.

4.2.3 Industrially produced artificial patina (examples)

Producers of copper sheets offer sheets with a layer of artificial patination – dark and green – too. Patinas are made continually on a production line by mechanical-chemical heat process or by application of gel containing copper nitrate and basic sulfates or copper chlorides. Extensive changes of patina composition can be expected to occur during its aging.

TECU®-Patina made by KM company (Germany) was used for cladding and roofing on both modern and historical buildings in Europe [21]. Sheets with Nordic Green patina made by Outocumpu company (Finland) were used on many historical and modern buildings around Scandinavia [19]. In the Czech Republic, both types of artificial patinas has been used on private or commercial buildings; artificial patina Nordic Green was used in roofing of Belveder (Queen Anna's summer palace) in Prague.

Sheets of artificial patina Nordic Green have been exposed for a long term at the atmospheric test sites of SVUOM and the changes of patina layer are being observed. A gradual wash-out occurred during the exposure but after approximately four years the green surface began to spread again. There is a gradual formation of natural patina, which grows through the layer of artificial patina. Computer picture

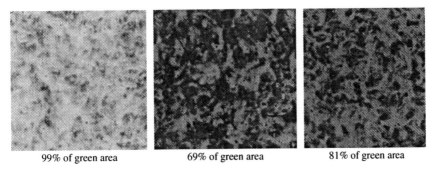

| 99% of green area | 69% of green area | 81% of green area |

Figure 2: Changes of the layer of artificial green patina Nordic Green.

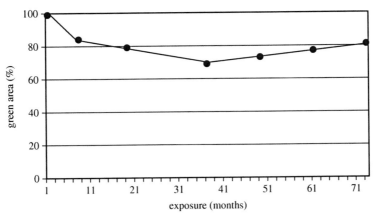

Figure 3: Changes of the layer artificial green patina Nordic Green during long-term exposure at a field test site Prague.

Table 12: X-ray diffraction analysis of patina layers.

Exposure	Time (years)				
	0	1	2	3	5
Composition of layer	Brochantite Posnjakite	Brochantite Posnjakite Cuprite	Brochantite Posnjakite Cuprite	Brochantite Posnjakite Cuprite	Brochantite Cuprite

analysis was used to evaluate these changes which allow for exact objective evaluation of the decreasing portion of the surface covered with green patina and the expression of gradual changes by numerical values (Figs 2 and 3). Transformation of the patina layer was observed at the same time (Table 12).

Table 13: Transformation of artificial patinas in various environments [13].

Patination agent	Composition of patina before exposure	Composition of patina after 2 years of exposure		
		Stockholm	Götteborg	Kopisty
Mixture of polysulfides	Very thin layer– unidentified	Cuprite Gerhardite	Cuprite Atacamite	Cuprite Posnjakite Brochantite
Mixture of polysulfides + $Cu(NO_3)_2$	Gerhardite	Gerhardite	Cuprite Atacamite Posnjakite Brochantite	Cuprite Posnjakite Brochantite Antlerite
$Cu(NO_3)_2$	Gerhardite	Cuprite Gerhardite	Cuprite Atacamite	Antlerite

Stockholm – low-pollution urban atmosphere influenced by traffic; Götteborg – urban atmosphere partially influenced by sea; Kopisty – industrial atmosphere, prevailing influence of SO_2.

4.2.4 Transformation of artificial patinas

Results of long-term tests prove that artificial patinas (produced in workshops or industrially) are transformed into natural patinas, which are in balance with the surrounding environment. The rate and degree of transformation depends on the type of atmosphere, its corrosivity and access of outdoor environmental impacts (Table 13) [22].

4.3 Conservation

Conservation of copper and bronze is used on historical objects and objects of art; it is used only rarely on construction elements (objects of arts on buildings). Conservation decreases the negative effect of outdoor atmosphere on copper and bronze surfaces and reduces the speed of deterioration of historical monuments. Conservation in the case of bronze and copper historical monuments means mainly treatment of surfaces by various types of waxes. Conservation should produce water-proof, hydrophobic, chemically stable, solid, elastic and even protective layer. Regular re-conservation at 1-2 years intervals can significantly reduce formation of patina on bronze and preserve original look of the material. Conservation agents are applied on objects covered with artificial green patina to protect it from wash-out and transformation.

Protective effect of waxes is the result of a barrier effect of the layer, mainly its low permeability for humidity and gaseous pollutants. Waxes provide only short-term protection (depending on type of wax, thickness of coating and corrosivity of atmosphere – 2 years at the maximum) and gradually grow old and loose elasticity and flexibility. They then need to be removed, and the surfaces must be re-conserved or treated with another protective system. Although very diluted waxes are used and the coatings formed are very thin, solvents and mechanical treatment have to be used to remove them.

Application of waxes and similar conservation agents changes the color of surfaces – mainly those covered with green patina, but partially also those covered with dark patina. Smallest changes of color are caused by beeswax.

Beeswax is the most often used traditional protective agent for copper and bronze sculptures. For application on historical objects beeswax is diluted with technical petrol. Beeswax applied by this way forms thin transparent coatings of 50–100 μm. Protective effect of beeswax results mainly from the barrier effect of its layer.

A very important property of beeswax for its use in conserving historical monuments is its water-repellent ability (hydrophobicity). Beeswax has a number of other advantages for protection of bronze monuments:

- very little changes of look, color and character of surfaces and patinas are observed;
- can be easily removed;
- does not change its look during gradual degradation and has no influence on the look of a monument;
- its protective effect is well known and verified by long-term use in practice.

The disadvantage of beeswax is a relatively short protective effect in outdoor atmospheres. The effectiveness of beeswax, especially its hydrophobic properties, is lowered by the impact of light (esteric bonds are being destroyed).

Microcrystalline waxes are mixtures of carbohydrates with linear, branched or cyclic chains. Waxes are esters of aliphatic acids C_{20}–C_{30} with high-order monofunction alcohols. Waxes are produced by cooling melted waxes in suitable solvents to form a structure of very small crystals, close to amorphous structure. These microcrystalline waxes can be applied in thin layers and produce smooth and bright coatings after polishing. Microcrystalline waxes are not as hydrophobic as beeswax. The protective effect of microcrystalline wax itself does not last longer than one year. Re-conservation is more difficult than in the case of beeswax layers. Gradual degradation creates an ugly-looking layer that should be removed before re-conservation. A number of commercially produced conservation waxes is based on microcrystalline waxes. Higher effectiveness of commercially produced conservation waxes is a result of the corrosion inhibitors contained by them.

Application of waxes coated at high temperatures is preferred and tested in Germany [23] and Sweden. Wax applied at high temperatures was also used in the Czech Republic during last three years. The protective effect of the conservation layer after approximately two years is satisfactory. Waxes applied in this way penetrate surface layers more easily and form thicker coatings in most cases.

Application of organic coatings with prevailing barrier effect seems to be effective only in few cases, one of the reasons being less suitable aesthetic effect after application on bronze monuments. General repair of old coatings, especially removing them, is hard and almost impossible in case of applying coatings over patina.

The problem of coatings is tried to solve them by new hybrid organic-inorganic coatings, which are applied in very thin layers. International task EUR 16637 EN to develop Ormocer coating was finished in 1995 [24]. The coating has not been sufficiently tested in practice so far. The aesthetic effect of the new coating may not be the most suitable one in all cases (high brightness, change of color of the base when applying the coating over green patina).

Fixation and anti-graffiti coatings on copper sheet semiproducts covered with industrially produced artificial patina can also be put into this category of protective measures of temporary effect.

5 Economic impacts

The process of atmospheric corrosion causes permanent damage in materials and objects. Considerable sums have to be spent on the maintenance or replacement of elements or damaged parts of objects. Economic considerations are highly influenced by the relation between the price and service life of materials/elements. Costs related to replacements need to be considered too.

It is very difficult to give statements about the economy of maintenance and conservation or restoration of sculptures, historical objects and objects of art. Prices differ a lot; they are often set individually; their service life is considered in time periods of centuries; and there are also specific relations between service life expressed as stability and service life considered in different aspects. The process of atmospheric deterioration of an object is often stopped by making a copy of an object and placing the original in a depository.

There are, therefore, only present examples of direct costs for repairs of copper elements of buildings. Data need to be understood rather in their relations than in terms of direct values, which differ both in time and place (Tables 14 and 15). Prices shown below include costs for categories other than direct material costs:

- price of material/element,
- design and technological preparation,
- amount of work, including its organization, and
- storage and transport of waste.

Comparing data in both tables leads to the conclusion that using copper for constructional and architectural purposes is economical. Although the material is relatively expensive, the service life of elements is long and the requirements for maintenance are small. Long service life is supported also by the fact that copper elements and bigger areas (roofing) are used mainly for offices, banks or important city buildings where professional treatment and proper maintenance can be expected.

If maintenance and replacement of copper elements is carried out on historical objects, costs are higher, usually by multiples of ten times.

Transfer of copper into the environment as a result of atmospheric corrosion, which is an ecological burden especially for water and soil, is considered a negative effect having economic consequences. Green corrosion products of copper and its

Table 14: Prices for repair and maintenance of roofing materials in the CR (in 1996) [25, 26].

Materials	Repair type	Repair cost ($€/m^2$)
Galvanized steel	Replacement	14.1
	Removal	0.5
Aluminum	Replacement	12.9
	Removal	0.7
Copper	Replacement	74.6
	Removal	0.5
Concrete tiles	Replacement	18.9
	Removal	2.3
Clay tiles	Replacement	12.6
	Removal	1.1
Special clay tiles	Replacement	24.6
	Removal	1.8
Slate	Replacement	46.0
	Removal	2.0

alloys are washed away damage also neighboring areas of construction materials. It is very difficult and expensive to remove traces of this effect and so it is carried out only exceptionally.

6 Impact of atmospheric corrosion of copper and its alloys on the environment

As a result of weathering of roofs, facades and other parts of constructions, which is accelerated due to acidifying pollutants, a significant part of metals is emitted to the biosphere. The main accumulation of metals occurs in urban areas where the influx of metals is the greatest. Elevated levels of heavy metals in the sludge used as fertilizers, in bottom sediments or in potable water may in the long run have adverse effects on biological systems and human health. Gravimetric measurement of exposed samples was the methodology used for assessment of release of heavy metals due to corrosion of materials within the UN ECE International Co-operative Programme on *Effect on Materials including Historic and Cultural Monuments* [27].

A layer of corrosion products is formed on surfaces of metals and thickness of metal or metallic layer goes on decreasing. Part of the corrosion product remains on the surface and other parts are washed away by precipitation or flown into the environment by wind as soluble salts or particles. Results of copper exposure show that a significant part of the corroded metal remains on the surface as a part of corrosion products: for example, as green-colored patina on copper roofs or

Table 15: Material corrosion costs by type of building [25] in thousands NOK06 (1995).

Type of building	Material										Total	Percentage
	Galvanized steel			Aluminum	Copper	Wood	Plaster	Concrete	Roofing felt	Brick		
	Untreated	Wire, profile	Treated									
Small house	2 616	0	3 663	136	63	27 030	3 896	2 869	2 060	317	42 650	21.6
Apartment block	5 169	0	9 434	547	97	4 636	16 994	7 307	2 319	597	47 101	23.8
Manufacture	1 055	0	1 260	109	31	326	321	387	1 212	71	4 771	2.4
Office	10 620	0	10 042	463	103	11 240	4 169	2 148	7 175	415	46 375	23.5
Hotels	91	0	98	5	1	110	41	42	70	8	465	0.2
Services	1 457	0	1 430	66	14	1 600	594	474	1 021	91	6 748	3.4
Agriculture	2 267	0	2 219	123	0	5 850	116	163	23	40	10 803	5.5
Other	1 711	0	1 433	65	23	2 760	1 212	827	701	84	8 815	4.5
Infrastructure	0	3 452	0	0	0	0	0	0	0	0	30 005	15.2
Total	24 983	3 452	26 552	1 515	332	53 552	27 342	14 216	14 581	1 623	197 732	100.0
Percentage	12.6	1.8	28.3	0.8	0.2	27.1	13.8	7.2	7.4	0.8	100.0	

Table 16: Ranges of copper corrosion rates and metal release rates based on results on corrosion attack from ICP Materials during the period 1987–95.

Exposure time	Corrosion rate[a] (g/m^2 a)	Metal release rate[a] (g/m^2 a)	Metal release (%)
1	6.7–13.4	0.3–1.2	3–21
2	5.2–10.3	0.5–1.6	7–33
4	5.0–9.3	0.7–1.8	10–33
8	3.8–6.8	0.8–2.0	20–42

[a]Average yearly value during the measuring period.

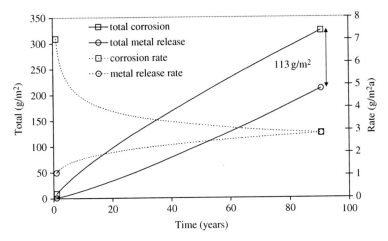

Figure 4: Calculated copper corrosion and metal release amounts and rates.

bronze sculptures. Results of average corrosion rates and metal release rates are shown in Table 16 after 1, 2, 4 and 8 years of exposure. Metal release rate is always lower than corrosion rate and the proportion increases with time due to decreasing corrosion rates and increasing metal release rates.

In constant environmental conditions, corrosion rate and metal release rate eventually reach the same value, corresponding to the so-called steady state where the composition and amount of corrosion products do not change with time. In Fig. 4 calculated copper corrosion and metal release amounts and rates as a function of exposure time based on dose-response functions and constant environmental conditions ($[SO_2] = 10\,\mu g/m^3$, $[O_3] = 40\,\mu g/m^3$, RH $= 75\%$, $T = 10°C$, Rain $= 700$ mm/a, and pH $= 4.5$) are illustrated. The difference between the amounts of corrosion and metal release (113 g/m^2 at steady state) corresponds to the amount of metal retained on the surface in corrosion products.

7 Case studies

During the period 1992–98 in compliance with EU 316 COPAL, inspections on about 200 copper and bronze objects have been performed by SVUOM specialists. Results are summarized in a database. Following are three examples:

- Copper decoration element from the roof of Museum of Applied Arts, Prague,
- Bronze statue of St. John of Nepomuk, Charles bridge, Prague, and
- Copper roof of Queen Anna's Summer Palace, Royal Garden, Prague.

All examples document the stage of non-treated copper and bronze cultural objects exposed over a very long period in very polluted urban areas.

7.1 Case study 1 – Decoration element on the roof of Museum of Applied Arts in Prague

Material: copper sheet, partly gilded, sheet remaining thickness 0.43–0.60 mm
Exposure: 100 years in urban atmosphere with corrosivity C4–C5
Defects:
- isolated cracks in area of forming and joints
- release of joints, non-functional joining by steel nails
- bimetallic corrosion
- occurrence of black spots in the area of soldered joints
- occurrence of crusts
- degradation of gilded surface

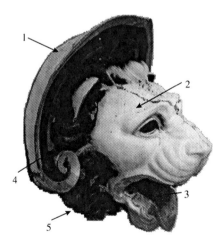 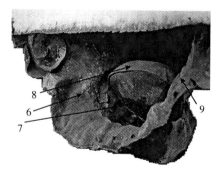

X-ray analysis of surface layer

Area of removal of samples	Appearance of layer	Composition
1 Upper hem	Bright green, turquoise	Brochantite
2 Upper edge of eye	Bright turquoise	Brochantite, antlerite
3 Mouth	Grey and green	Antlerite, quartz, gypsum
4 Profiled part of hem	Dark gray	Cuprite, antlerite
5 Lower dark hem	Red and brown	Cuprite, antlerite
6 Lower part of inside surface with dust and crusts	Bright turquoise with gray parts to dark gray and green	Antlerite, chalcantite, gypsum, $(NH_4)_2Cu(SO_4)_2 \cdot 6H_2O$
7 Lower part of inside surface	Turquoise, gray parts	Chalcantite, antlerite, brochantite, gypsum
8 Upper part of inside surface	Grey-and-green to dark green, dark gray	Antlerite, quartz
9 Inside surface of hem	Brown, gray and green	Cuprite, quartz, antlerite

Microprobe analysis: chlorides in surface layers (samples 4 and 5)
Metallography cross: uniform microstructure of copper with patina layer

7.2 Case study 2 – Sculpture of St. John of Nepomuk on the Charles bridge in Prague

Material: bronze alloy, gilded steel decorative elements
Exposure: 300 years in urban atmosphere with corrosivity C4
Defects:
- small holes on the most open areas (forehead, cap)
- dark surface with dust deposits in shelter position
- corrosion on the bonders of grain of alloy
- application of lead as filling material for holes and openings

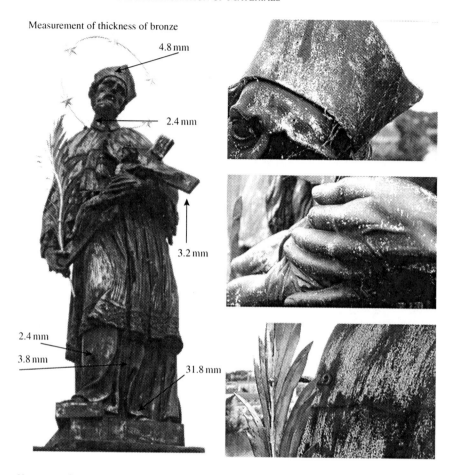

Measurement of thickness of bronze

4.8 mm

2.4 mm

3.2 mm

2.4 mm

3.8 mm

31.8 mm

X-ray analysis of surface layer

Area of removal of samples	Appearance of layer	Composition
Upper surface of cap	Green-and gray	Brochantite, cuprite
Lower part of coat, shelter	Black	$(NH_4)_2Cu(SO_4)_2 \cdot 6H_2O$, graffite
Open surface of body	Green	Brochantite, quartz, cuprite, anglesite $(PbSO_4)$

7.3 Case study 3 – Roof of Queen Anne's Summer Palace in Royal Garden in Prague

Material: copper sheet, sheet remaining thickness 0.80 mm on the most of roof
sheet remaining thickness 0.35–0.40 mm on the east saddle part

Exposure: 100–400 years in urban atmosphere with corrosivity C4–C5
Defects:
- cracks and perforations on the sheets with minimum thickness
- occurrence of black spots on sheets

East side of roof

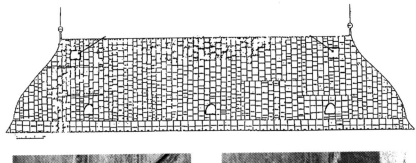

X-ray analysis of surface layer

Area of removal of samples	Appearance of layer	Composition
1 Cover of window	Middle green	Brochantite
2 Sheets on east side	Yellow and green	Brochantite, quartz, gypsum
3 Area with black spots	Black	Cuprite

Microprobe analysis of spots: high amount of Fe
Metallography cross: layered patina – lower layer cuprite; upper layer brochantite, antlerite pits

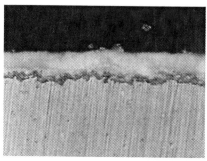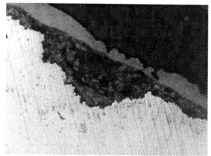

200×

Reconstruction:
- realized in 2000–03
- replacing of damaged sheets by new copper sheets
- application of artificial green patina on replaced sheets

References

[1]   Knotkova, D., Kreislova, K., Vlckova, J. & Had, J., *Effect of Atmospheric Environments on Copper and Bronze Cultural Monuments*, COPAL I–VI, SVUOM technical reports to project EU 316 EUROCARE COPAL, 1994–1999.

[2]   Stöckle, B., Krätschmer, A., Mach, M. & Snethlage, R., *Corrosion Attack on Copper and Cast Bronze. Evaluation after 8 Years of Exposure*, Report No 23 of UN ECE ICP on Effects on Materials Including Historic and Cultural Monuments, May 1998.

[3]   Kucera, V., Tidblad, J., Faller, M. & Singer, B., *Release of Heavy Metals due to Corrosion of Materials*, Report of UN ECE ICP on Effects on Materials Including Historic and Cultural Monuments, 2001.

[4]   Mattsson, E. & Holm, R., Atmospheric corrosion of copper and its alloys, *Atmospheric Corrosion*, ed. W.H. Ailor, John Wiley and Sons: New York, NY, pp. 365–381, 1982.

[5]   Graedel, T., Nassau, K. & Franey, J.P., Copper patinas formed in the atmosphere – I. Introduction. *Corrosion Science*, **27**(7), pp. 639–649, 1987.

[6]   Graedel, T., Copper patinas formed in the atmosphere – II. A qualitative assessment of mechanisms. *Corrosion Science*, **27**(7), pp. 721–740, 1987.

[7]   Graedel, T., Copper patinas formed in the atmosphere – III. A semi-quantitative assessment of rates and constraints in the greater New York metropolitan area. *Corrosion Science*, **27**(7), pp. 741–769, 1987.

[8]   *Proc. of ICOMOS Conf. 'Metallrestaurierung'*, Mnichov, 23–25 October 1997, ed. Martin Mach, Arbeitshefte des bayerixchen Landesaimtes fur Denkmalpflege, Band 94, Mnichov, 1998.

[9]   Knotkova, D., Vlckova, J., Kreislova, K. & Had, J., Atmospheric corrosion of bronze and copper statues, roofs and cladding on Prague's monuments, *Proc. of Art '96*, Budapest, October 1996.

[10] Had, J. & Knotkova, D., Patiny a korozní produkty medi a medenych slitin. *Zpravy pamatkove pece*, **60**(1), pp. 8–10, 2000.

[11] *Proc. of Workshops of Project Eurocare – EU 316 Copal*, Praha, 1995, and Wien, 1996, Institut fur Silikatchemie und Archeometrie: Wien.

[12] Franey, J.P. & Davis, M.E., Metallographic studies of the copper patina formed in the atmosphere. *Dialogue/89 – The Conservation of Bronze Sculpture in the Outdoor Environment: A Dialogue Among Conservators, Curators, Environmental Scientists, and Corrosion Engineers*, ed. T. Drayman-Weisser, NACE: Houston, TX, 1992.

[13] Riederer, J., The present level of techniques for conservation of outdoor bronze sculptures. *Proc. of Copal EU 316 Workshop*, Budapest, May 1994, and Vienna, 1995.

[14] Kreislova, K., Knotkova, D. & Had, J., Database of corrosion damage of copper and bronze monuments in Prague. *Proc. of the 5th International Congress on Restoration of Architectural Heritage*, Florence, Italy, 17–24 October 2000.

[15] Ashurst, J. & Ashurst, N., *Practical Building Conservation English Heritage Technical Handbook*, Volume 4: Metals, Gower Technical Press Ltd.: Aldershot, UK, 1988.

[16] Gullman, J. & Tornblom, M., *Bronze Sculpture. The Making and Unmaking*, ed. The Central Board of National Antiquities and National Historical Museum, Stockholm, 1994.

[17]   Pago, L., *Konzervovanie nezeleznych kovov v muzejnej praxi*, Archeologicky ústav, CSAV Brno, 1984.

[18]   Kuttainen, K., *Metal Losses During Cleaning and Patination of Cast Bronze*, Report SCI 1991-1, Stockholm.

[19]   Outokumpo copper Nordic Green – method for patination of new-laid copper. Instruction for use, technical specification by producer, 1996.

[20]   Hughes, R. & Rowe, M.(eds.), *The Colouring, Bronzing and Patination of Metals*, Thames and Hudson Ltd.: London, 1991.

[21]   Med v architekture, VUT Brno, technical specification by producer, 2000.

[22]   Johansson, E., *Korrosion samt restaurering och konservering av bronsstatyer*, Report SCI, Stockholm, 1995.

[23]   Meissner, B., Doktor, A. & Mach, M., Bronze- und Galvanoplastik. *Geschichte – Materialanalyse – Restaurierung*, Landesamt fur Denkmalpflege Sachsen, Mnichov 2000.

[24]   Römich H., (ed.), *New Conservation Methods for Outdoor Bronze Sculptures*, Final report of EUN 16637 EN Protection and conservation of European cultural heritage, 1995.

[25]   Hamilton, R. *et al.*, *Rationalized Economic Appraisal of Cultural Heritage (REACH)*, Second Annual Report and Deliverables, Project EU, May 2000.

[26]   Knotkova, D., Sochor, V. & Kreislova, K., *Evaluation of Damages to Materials and Buildings Caused by Pollution*, report SVUOM, Prague, 1997.

[27]   Singer, B., Woznik, E., Krätschmer, A. & Doktor, A., *Results from the Multipollutant Programme: Corrosion Attack on Copper and Bronze after 1 and 2 Years of Exposure*, Report No. 37 of UN ECE ICP on Effects on Materials Including Historic and Cultural Monuments, June 2001.

CHAPTER 5

Environmental deterioration of concrete

A.J. Boyd[1] & J. Skalny[2]
[1]*Department of Civil Engineering & Applied Mechanics, McGill University, Montreal, Quebec, Canada.*
[2]*Materials Service Life, Holmes Beach, Florida, USA.*

1 Introduction

Concrete is a man-made composite material produced in its simplest form from Portland *cement*, coarse and fine aggregates, and water. In modern times, concrete may contain supplementary cementing materials (such as fly ash, blast furnace slag, or silica fume), various fillers (e.g. limestone), chemical admixtures (i.e. set retarders, set accelerators, water reducers, corrosion inhibitors), and a variety of fibers.

1.1 Physics and chemistry of concrete formation

Portland cement and other Portland cement clinker-based cements are the most common cementing materials used in concrete production. The total world production of cement is estimated to be above 1.7 billion metric tons per year, enabling the production of more than 10 billion metric tons of concrete annually.

Cement clinker is produced in high-temperature heat exchangers (called cement kilns) from a ground raw mix composed of calcium carbonate and siliceous, aluminous, and ferric minerals, plus unavoidable minor components such as alkalis, sulfates, magnesia, etc. The burned and cooled clinker is an intimate mix of various *clinker minerals* and, when finely ground with a limited amount of calcium sulfate, provides the resulting cement with certain properties that become important when it reacts with water to produce the 'glue' that binds the aggregates and other concrete components into structural concrete (see Table 1, Fig. 1) [1].

Table 1: Components of Portland clinker-based concrete.

Concrete components	• Portland cement • Supplementary cementing materials (fly ash, granulated blast furnace slag, silica fume, etc.) • Coarse and fine aggregate (gravel, crushed stone, sand) • Water • Chemical admixtures • Fillers (e.g. limestone) • Fibers • Reinforcing steel
Cement components	• Clinker ground with a limited amount of calcium sulfate (gypsum, hemihydrate, etc.) and, in blended cements, with fly ash, slag, and/or other mineral components
Clinker components[1,2]	• Tricalcium silicate (C_3S or Ca_3SiO_5) • Dicalcium silicate (C_2S or Ca_2SiO_4) • Tricalcium aluminate C_3A or ($Ca_3Al_2O_5$) • Tetracalcium aluminate ferrite (C_4AF or $Ca_4Al_2Fe_2O_{10}$) • Alkali sulfates • CaO (limited amount) • MgO (limited amount)
Primary cement-water reaction products in hardened concrete[3]	• Calcium silicate hydrate (C-S-H) • Calcium hydroxide ($Ca(OH)_2$ or CH) • Ettringite (hexacalcium trisulfoaluminate hydrate or $\mathbf{C_6A\bar{S}_3H_{32}}$) • Monosulfate (tetracalcium monosulfoaluminate hydrate or $\mathbf{C_4A\bar{S}H_{12}}$) • Calcium carbonate, $CaCO_3$ (product of atmospheric carbonation)

[1] The clinker minerals may form solid solutions, thus the given chemical composition is only approximate.

[2] The cement/concrete literature nomenclature uses a shorthand approach to oxide chemistry, C for CaO, S for SiO_2, A for Al_2O_3, F for Fe_2O_3, $\mathbf{S}$ for SO_3, H for H_2O, M for MgO, K for K_2O, and N for Na_2O.

[3] All these hydration products play a certain role in improving or diminishing concrete durability.

Typical clinker oxide and mineralogical composition of cement, and the ASTM cement types, are given in Table 2. The types of cement are characterized by ASTM (American Society for Testing and Materials) standards as possessing certain properties, such as resistance to sulfates or early strength [1, 2]. Other standards, although

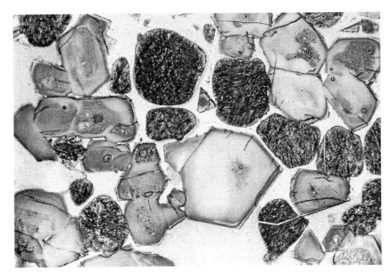

Figure 1: Portland cement clinker: angular crystals of alite (C_3S) and rounded crystals of belite (C_2S). Polished thin section, light optical microscopy, magnification $400\times$ (Courtesy: Portland Cement Association).

they may differ in detail from ASTM classification, characterize cement types by similar criteria. These characterizations are essentially related to the mineralogical composition of the particular cement type because the proportions of the various minerals present in clinker control to a certain degree the performance of the cement in concrete. For example, high early-strength cement will typically be made of clinker containing high amounts of fast-reacting tricalcium silicate and tricalcium aluminate, and will be ground to higher that usual fineness to increase the rates of the hydration reactions and thus, the rate of strength development. In a similar way, the so-called sulfate-resisting cements, such as ASTM Type V, will have low amounts of tricalcium aluminate to decrease their susceptibility to the formation of expansive ettringite in cases where the concrete is exposed to a sulfate-rich environment.

When cement is mixed with water, the compounds present in the cement react to form various hydration products. The process of hydration is complex, as the individual compounds hydrate exothermally at different rates, releasing different amounts of heat, forming hydration products of different crystalline structures, and having unique affects on the physical properties of concrete and its stability in the environment of use (Fig. 2).

Table 3 briefly summarizes the most important hydration reactions of Portland cement. The key reaction product of the hydration of calcium silicates is *calcium silicate hydrate* (C-S-H). It is a high specific surface area, semi-amorphous hydrate of variable composition that has the ability to bind together the other concrete components (sand, aggregate), thus forming an 'artificial stone'. The exact chemical

Table 2: Characteristic compositions of Portland cement.

Typical oxide composition of Portland clinkers (%)	CaO: 64–65 SiO$_2$: 20–22 Al$_2$O$_3$: 4–7 Fe$_2$O$_3$: 3–5 MgO: 1–4 SO$_3$: 0.3–2.5 Alkalis: 0.2–2.0
Typical compound composition of Portland cements (%)	Tricalcium silicate, alite: 36–72 Dicalcium silicate, belite: 18–32 Tricalcium aluminate: 0–14 Tetracalcium aluminate ferrite: 1–18
ASTM types of Portland cement	Type I: Normal Type II: Moderate sulfate resistant and heat of hydration Type III: High early strength Type IV: Low heat of hydration Type V: High sulfate resistance

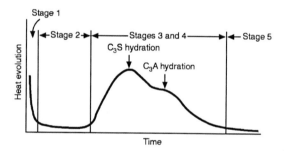

Figure 2: Heat evolution of typical Portland cement paste as a function of time. Stage 1 is related to the heat of wetting or initial hydrolysis (C$_3$A and C$_3$S hydration). Stage 2 represents a dormant or induction period related to the initial set. Stage 3 shows the accelerated reaction of the clinker minerals – primarily C$_3$S – that determines the rate of hardening and final set. Stage 4 shows the deceleration in the formation of hydration products and determines the rate of early strength gain. Stage 5 represents the slow, steady formation of hydration products establishing the rate of later strength gain.

composition of C-S-H depends on the cement composition, processing conditions, and the environment (temperature, humidity). *Calcium hydroxide* (Ca(OH)$_2$ or CH) is the primary by-product of calcium silicate hydration and its importance is primarily in maintaining a high pH of the hardened concrete system. Both C-S-H and

Table 3: The basic hydration reactions of the primary cement components (oversimplified).

Reactants	Reaction products	Importance
C_3S, H	C-S-H, CH	Rapid rate of hydration. Semi-amorphous C-S-H is the primary binder giving concrete its engineering properties. CH maintains high alkalinity, but is of marginal importance as a binder; may result in low durability.
C_2S, H	C-S-H, CH	As above. Slow rate of hydration.
C_3A, H, $CaSO_4 \cdot 2H_2O$	$C_6AS_3H_{32}$ (ettringite, hexacalcium aluminate trisulfate 32-hydrate)	C_3A hydration is rapid, but rate is controlled by calcium sulfate addition. Resulting ettringite may be expansive under certain conditions, thus causing volume instability.
C_3A, H, $C_6AS_3H_{32}$	C_4ASH_{12} (monosulfate, tetracalcium aluminate monosulfate 12-hydrate)	Monosulfate may under certain conditions form additional ettringite.

CH are vulnerable to external chemical influences, thus playing an important role in concrete durability.

The aluminate and alumino-ferrite phases react in the presence of sulfate to form ettringite or monosulfate. Sulfate is added to Portland cement to regulate the time of the initial and final set. In the absence of sulfate, the hydrating cement would solidify before the concrete is properly placed and finished. Thus, ettringite is a normal and expected product of early cement hydration and its presence in concrete is not, by itself, evidence of concrete deterioration. As will be explained later, ettringite formation at later ages, when the hardened concrete is exposed to an external source of sulfates, may be deleterious because, depending upon the environmental conditions, such ettringite may be expansive and may lead to severe cracking of the concrete microstructure.

During the hydration process, the overall volume of the solid-water system decreases slightly, thus leading to a possible volume change and cracking. An example of such a volume change is the so-called *plastic shrinkage*, which occurs while the concrete is still plastic and is a result of chemical shrinkage and evaporation of water from the concrete surface. The volume change of concrete depends on cement composition, amount of water used (w/c), and environmental conditions. The overall volume change of concrete is schematically explained in Fig. 3 [1].

The total porosity of concrete is highly dependent upon the original amount of water used in its production. This is usually expressed in the form of water-to-cement ratio, w/c (or w/cm), the amount of water per unit weight of cement. The typical range of w/c used is between 0.35 and 0.60. Because the amount of water needed for complete hydration of cement is much lower than that needed to produce a workable mix, in virtually all instances, the porosity of the resulting concrete is higher than desired. This has important implications with respect to durability.

Because the density of the hydration products is lower than that of the anhydrous solids present in clinker, the porosity of the system decreases as hydration of the clinker minerals progresses. As mentioned previously, even at complete hydration the paste will have a certain amount of porosity as a result of excess mix water. This proportion is increased further because the formed hydration products possess some intrinsic porosity of their own and air is entrapped during processing. At very low overall porosities, the hardened material has a *closed* pore structure, i.e. the pores are not interconnected. At higher porosities, the pores become interconnected and form an *open* pore structure. The degree of porosity thus influences concrete permeability.

Strength of concrete, whether compressive or bending, is also indirectly proportional to w/c ratio, as porosity severely diminishes the strength of all materials. However, the relationships are complex and one cannot assume a direct and simple correlation between w/c, porosity, and strength. A typical w/c versus compressive strength relationship is presented in Fig. 4. Note the rapidly decreasing strength with increasing w/c and the wide range of strengths that are possible for a given w/c. Such variability is caused by the cement quality (type of cement, its rate of strength development), the mix design (type and amount of aggregate, use of admixtures and supplementary cementing materials), the processing and curing conditions

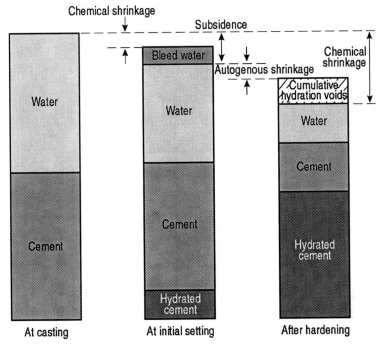

Figure 3: Volumetric relationship between subsidence, bleed water, chemical shrinkage, and autogenous shrinkage (after set) during cement hydration. Not to scale. (Courtesy: Portland Cement Association). *Chemical shrinkage* represents the overall reduction in volume of the solids plus liquids in the cement paste during hydration. *Autogenous shrinkage* is the visible dimensional change caused by hydration; it is driven by chemical shrinkage but, because of the rigidity of the system, is lesser than the total (chemical) shrinkage. *Subsidence* is the vertical shrinkage of the material before initial set and is caused by settlement of the solids (also referred to as *bleeding*).

(temperature, humidity, mixing procedure, timing of process conditions), and by the quality of workmanship (often uncontrolled!).

It must be emphasized that while strength is the most important concrete property from a structural point of view, it does not automatically imply adequate durability in a given environment. It is entirely possible to produce concrete that reaches the required strength but is vulnerable to severe chemical deterioration. Therefore, *strength and durability* (and other expected qualities) must be taken into consideration during concrete mix proportioning, structural design, processing, and maintenance. Designing concrete without taking its environmental stability into consideration is an unsound practice and may well be economically disastrous over the service life of the structure.

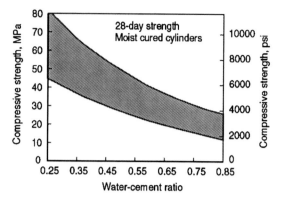

Figure 4: Compressive strength versus water-to-cement ratio (w/c) of concrete after 28 days of laboratory moist curing (Courtesy: Portland Cement Association).

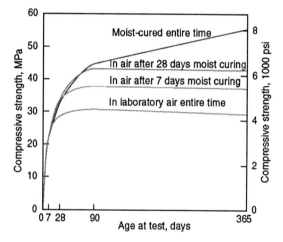

Figure 5: Effect of air versus moist curing on the compressive strengths of laboratory-cured concrete (Courtesy: Portland Cement Association).

Curing of concrete is the most important, though often misunderstood or neglected, part of the concrete-making process. It is during curing that the cement components hydrate, form hydration products of certain chemical and microstructural qualities within a prescribed space. It is the hydration process that leads to development of such important qualities as water-tightness, strength, modulus of elasticity, durability in given environments, etc. Examples of the importance of curing are shown in Figs 5 and 6. Note not only the relationship of strength to the form of curing, but also the inverse relationship of early and late strength as a function of curing temperature. Both these facts must be taken into consideration in engineering practice and both have serious economic implications.

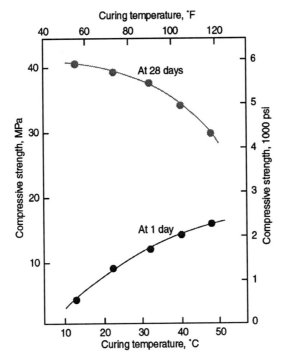

Figure 6: Effect of curing temperature on the early and late compressive strength of laboratory-cured concrete (Courtesy: Portland Cement Association).

For additional information on the chemical and microstructural aspects of the transformation of the cement-water-aggregate system into structural concrete, please refer to the available literature [1, 3, 4].

1.2 Mechanisms of concrete deterioration

All concrete structures are exposed to the environment. Thus, concrete mixture proportioning, structural design, processing, and maintenance have to take into consideration the environmental conditions under which the concrete is to function during its expected service life. Because the environments in which concrete may be used vary widely (cold to warm, dry to wet, repeated temperature and humidity changes, seawater, aggressive soils, etc.), it is a mistake to assume that any concrete will be stable under all conditions. This basic concept is often neglected, thus resulting in premature deterioration of structural concrete components and products. Examples of deteriorated concrete are shown in Figs 7–9.

The mechanisms of concrete deterioration may be chemical, physical, or mechanical in nature, but the phenomena are often interrelated. Typical examples of such mechanisms are sulfate attack, freezing and thawing, and abrasion, respectively.

Figure 7: Surface of a concrete slab damaged by exposure to sulfate containing soils (Photo: J. Skalny).

Figure 8: Concrete railroad sleeper damaged by combined action of alkali–silica reaction (ASR) and excessive heating resulting in internal sulfate attack (Photo: J. Skalny).

A brief overview of the most typical mechanisms of deterioration of concrete components is given in Table 4.

According to the ACI Guide to Durable Concrete [5], durability of concrete is defined as its 'ability to resist weathering action, chemical attack, abrasion or

Figure 9: Concrete pier showing signs of reinforcement corrosion (Courtesy: N. Berke).

Table 4: Primary mechanisms of concrete deterioration.

Mechanical deterioration of the concrete surface	• Abrasion • Erosion • Cavitation
Deterioration of the reinforcement Deterioration of the aggregate	• Corrosion of steel • Freezing and thawing • Alkali–aggregate reactions
Deterioration of the cement-based matrix	• Freezing and thawing • Sulfate attack (internal, external) • Seawater attack • Acid attack • Carbonation • Salt crystallization

any other process of deterioration. Durable concrete will retain its original form, quality, and serviceability when exposed to its environment.' Expressed somewhat differently, a concrete product or structure has to retain its quality and serviceability in the environment of use throughout its expected service life.

It needs to be noted that, in addition to other preconditions, most deterioration mechanisms require the presence of high humidity or liquid water; dry chemicals rarely attack concrete. It should also be pointed out that multiple deterioration mechanisms can occur simultaneously (e.g. carbonation and corrosion, alkali–silica reaction and sulfate attack, sulfate attack and salt crystallization).

The required concrete 'quality' may differ from one structure and one kind of environment to another. In some cases, strength and modulus of elasticity may be the only crucial properties, whereas under other conditions permeability and resistance to aggressive external chemicals may be critical in addition to structural integrity. If economy is also taken into consideration, it is clear that concrete should be designed for a purpose. The available materials knowledge should be implemented in the design phase during which the desired properties and property levels are planned and recommended. Such an approach is often ignored, resulting in over- or under-designed concrete properties for a particular application.

The maximum expected service life of a concrete product or structure varies, based on the chemical and atmospheric environment the concrete is exposed to, and clearly depends on the original design, the actual concrete quality, and the level of maintenance.

2 Chemical deterioration of cement paste components

Like the processes leading to the development of concrete properties, the processes causing premature deterioration of the concrete matrix are usually chemical in nature as well. This fact is often unrecognized or ignored. The chemical processes of deterioration may have physical (volume change, porosity and permeability alteration) or mechanical (cracking, strength loss) consequences, or both. Typical examples are acid attack, alkali–silica reaction, carbonation, and sulfate attack.

The most vulnerable part of the concrete composite is the cement paste – the 'glue' binding the aggregate together – as it is the 'first line of defense' directly exposed to the environment. *Cement paste* (or concrete matrix) is composed of the products of the cement-water reaction (hydration products), of unreacted cement particles, air voids of various sizes, and porosity. The origin of the larger air voids (entrapped air) is usually the concrete processing techniques used (mixing, placing), while the smaller air voids are generally the result of purposeful air entrainment used to decrease the susceptibility of concrete to freezing. Porosity is the result of hydration wherein part of the mix water that was consumed during the hydration reactions leaves behind empty, more or less interconnected porosity. Another source of porosity is microcracks caused by several possible mechanisms of volume change. The voids and other forms of porosity, depending upon the relative humidity, are partially water-filled.

Porosity is one of the most important properties of hardened concrete, as its magnitude and form controls the permeability of concrete with respect to water vapor and water-soluble ionic species. Most applications of concrete assume that it has certain 'barrier' properties protecting it from the environment, therefore control of porosity and permeability is crucial. The most important methods of permeability control are proper mix proportioning and effective curing [1].

The porosity of cement paste depends on the amount of mix water used, usually expressed as water-to-cement ratio (w/c) or water-to-cementing materials-ratio (w/cm). Since the complete hydration of cement requires only a specific amount of water, any water beyond this theoretically quantity will result in increased porosity. Thus, higher w/c ratios will result in higher the porosity concretes. At w/c ratios below approximately 0.4 (4 parts of water per 10 parts of cement), the individual pores are separated, thus the permeability is low. However, at a w/c ratio above 0.4, the porosity becomes interconnected, leading to increased permeability with further increases in w/c [3, 6]. Permeable concrete is not durable in aggressive chemical environments.

Proper curing (time-temperature-humidity combination) controls the degree to which the cement present in the original mix reacts with water to form hardened cement paste. Curing also influences the quality and nature of the microstructure formed. Because the specific density of the hydration products is lower than that of the anhydrous cement (i.e. their volume is larger than that of the cement), hydration reactions lead to an overall decrease in porosity. At a given w/c ratio, the porosity, and thus the permeability, of a well-cured concrete is lower than that of a poorly cured concrete. For illustration of the utmost importance of water-to-cement ratio and proper curing, see Figs 10 [3] and 11 [1].

It is obvious from the above discussion that the kinetics of chemical degradation of concrete is controlled by the diffusion of ions into, out of, and through the concrete pore structure. The mechanisms of ionic diffusion in both saturated and unsaturated cementitious systems are complex as they involve interrelated phenomena such as transport of fluids and ionic species, as well as physical and chemical interactions of ions within the solid phase [7].

2.1 Aggressive soils and ground water

The most widely spread concrete damage by chemicals present in soil and groundwater is attack by sulfates. Therefore, most of this chapter will be dedicated to various forms of sulfate attack.

However, soil and groundwater may contain other chemicals that can affect concrete durability, though in a much less damaging way than sulfates. Soluble chlorides, carbonates, pyrite, hydrogen sulfide, phosphates, and other natural soluble species may react with concrete paste components, possibly leading to increased porosity, decomposition of cement paste components, formation of new chemicals, volume changes, or a combination of these factors. Such reactions may be important to consider when multiple damage mechanisms are operable. As in the case of sulfate attack, proper mixture proportioning (low concrete porosity and permeability), effective curing, and separation of the structure from the source of the soluble chemicals involved are the primary remedies for avoiding deterioration problems.

2.1.1 Sulfate attack
Sulfate attack is a generic name for a series of interrelated, sometimes sequential, chemical and physical processes leading to recrystallization or decomposition

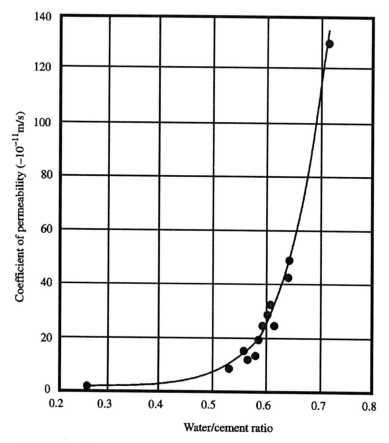

Figure 10: Relationship between water-to-cement ratio and coefficient of permeability. Note the rapidly increasing permeability above about w/c = 0.4; related to interconnected porosity.

of hydration products and deterioration of the concrete matrix microstructure, subsequently inducing degradation of the expected concrete properties, including strength, volume stability, and durability [8, 9]. The source of sulfates may be internal, originating from the concrete components (cement, aggregate, etc.), or external, such as exposure to sulfate-bearing soils, groundwater, or industrial and agricultural waste.

2.1.1.1 Internal sulfate attack   This attack may be caused by (a) out-of-standard amounts of sulfates in the cement or aggregate (*composition-induced internal sulfate attack*) or (b) improper processing of the concrete, primarily by excessive curing temperature (*heat-induced internal sulfate attack*). Heat-induced sulfate attack is often referred to in the literature as *delayed ettringite formation* or DEF [10–12].

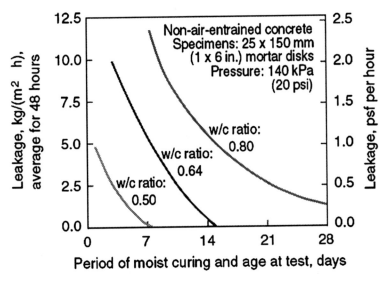

Figure 11: Effect of w/c and curing duration on permeability of cement mortar. Note that leakage is reduced as the w/c is decreased and the curing period is increased [1].

As a result of limitations imposed by national and international standards, the total clinker sulfate content in modern cements and the sulfate present in aggregate are well controlled and, consequently, *composition-induced internal sulfate attack* is uncommon.

Heat-induced internal sulfate attack (DEF) is a phenomenon that began to draw attention in the early 1990s as a result of severe durability problems observed with heat-cured concrete railroad sleepers. After years of scientific debate and disagreement, the mechanistic issues are now believed to be mostly resolved [11]. The mechanism of damage can be briefly described as follows:

When freshly mixed concrete is cured at high temperatures, the usual early formation of ettringite does not occur. The absence of ettringite after heat curing is not the result of decomposition of the previously formed ettringite, but the consequence of the redistribution of aluminate and sulfate into other phases – specifically C-S-H, AFm (aluminate-ferrite monosulfate hydrate), and hydrogarnet. The thermodynamic stability and solid solution range of the two latter compounds is not well known and is an obstacle to a more complete understanding of the system as a whole. Upon subsequent cooling to ambient temperature, supersaturation with respect to ettringite enables the previously formed monosulfate to react with sulfate and aluminate ions present in the system (C-S-H, pore solution, etc.) and form expansive *primary* ettringite within the C-S-H. As a consequence, the hydrated paste expands and forms gaps around the (nonexpansive) aggregate particles. In severe cases, this may lead to overall expansion and cracking of the concrete element or product (see Fig. 8). With time, the submicroscopic ettringite particles recrystallize

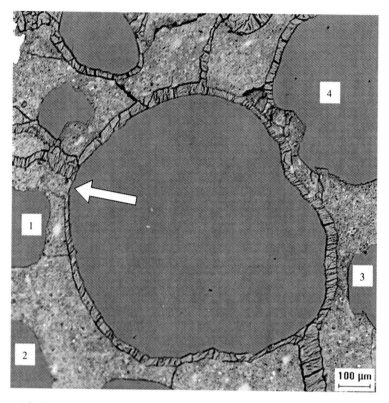

Figure 12: Secondary ettringite in gaps around the aggregate particles in a mortar exposed to elevated temperature. SEM backscattered image (Courtesy: S. Diamond).

into the open spaces of the paste (gaps, cracks, small pores) as larger crystals; such *secondary* ettringite is easily detectable by light optical or electron microscopy (see Fig. 12) [13].

2.1.1.2 External sulfate attack   This attack is induced by penetration of sulfates into concrete structures exposed to sulfate-rich environment, primarily high-sulfate soil or groundwater. The most common forms of sulfate in soil are calcium sulfate dihydrate (gypsum), magnesium sulfate, or sodium sulfate. Depending on the environmental conditions (water access, groundwater movement, changes in temperature and humidity, etc.) and concrete quality (primarily permeability, degree of cracking), the damage to concrete can be severe [5, 8, 9].

Sulfate attack on concrete results in chemical and microstructural changes to the hardened cement paste matrix, and these changes eventually lead to degradation of the physical and mechanical properties expected of any particular structure. Properties that might be affected include the barrier properties (permeability, water

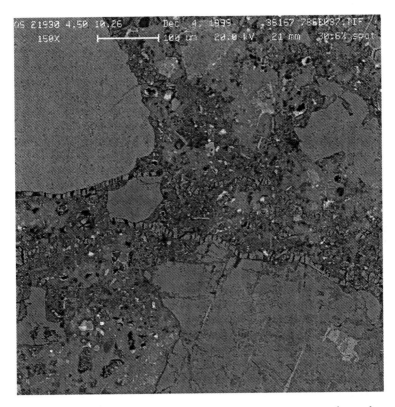

Figure 13: Deposits of ettringite in porous concrete paste matrix and paste–
aggregate interfaces. Backscattered SEM image (Courtesy: P.W.
Brown).

tightness), modulus of elasticity, and strengths, particularly bending strength. Typically observed damage includes deterioration of the evaporative surfaces (sulfate salt crystallization), softening of the matrix (gypsum formation, C-S-H decomposition), and volume instability of the concrete (expansive ettringite formation). However, as stated earlier, the presence of ettringite in concrete is not in itself a sign of sulfate attack. Because of its complexity, it is difficult to assess sulfate attack by only one of the many possible reactions between sulfate and cement paste components, and many standard tests are less than adequate in sulfate attack characterization. A multidisciplinary approach is usually needed to explain the damage phenomena in detail.

The micrograph in Fig. 13 represents a hardened cement paste containing ettringite in excess of that formed during the hydration of cement, a result of sulfate solution penetration into a permeable concrete matrix. An overview of the various sulfate-related mechanisms of deterioration is given in Table 5 [14].

Sulfate attack-related expansion and cracking is usually attributed to the formation of expansive ettringite through the reaction of external sulfates with clinker

Table 5: Reactions in Portland cement-based systems involving sulfates.[1]

'Physical' sulfate attack

$$2Na^+ + SO_4^{2-} \quad \rightarrow \quad Na_2SO_4 \cdot 10H_2O$$
(solution)   (evaporation)   (solid)

$$Na_2SO_4 \cdot 10H_2O \quad \leftrightarrow \quad Na_2SO_4$$
(mirabilite)   (repeated recrystallization)   (thenardite)

Pyrite oxidation (leading to combined sulfate and sulfuric acid attack)

$$2FeS_2 + 7O_2 + 2H_2O \rightarrow 2FeSO_4 + 2H_2SO_4$$
(pyrite)

or

$$4FeS_2 + 15O_2 + 2H_2O \rightarrow 2Fe_2(SO_4)_3 + 2H_2SO_4$$

Ettringite formation

$$C_3A + 3C\bar{S}H_2 + 26H \quad \rightarrow \quad C_6A\bar{S}_3H_{32}$$
(gypsum)   (ettringite)

$$C_4A\bar{S}H_{12} + 2C\bar{S}H_2 + 16H \quad \rightarrow \quad C_6A\bar{S}_3H_{32}$$
(monosulfate)

$$C_4A_3\bar{S} + 8C\bar{S}H_2 + 6CH + 74H \rightarrow 3C_6A\bar{S}_3H_{32}$$
(4-calcium 3-aluminate sulfate)

Gypsum formation

$$2OH^- + SO_4^{2-} + Ca^{2+} \rightarrow C\bar{S}H_2 \quad or \quad C\text{-}S\text{-}H + SO_4^{2-} \rightarrow C\bar{S}H_2$$
(from calcium hydroxide)   (from calcium silicate hydrate)

$$2Na^+ + SO_4^{2-} + Ca^{2+} + aq. \rightarrow 2Na^+ + C\bar{S}H_2 + aq.$$
(in the presence of sodium sulfate)

Thaumasite formation

$$3Ca^{2+} + SiO_3^{2-} + CO_3^{2-} + SO_4^{2-} + 15H_2O \rightarrow 3CaO \cdot SiO_2 \cdot CO_2 \cdot SO_3 \cdot 15H_2O$$
(thaumasite)

Reactions involving magnesium sulfate

$$Mg^{2+} + SO_4^{2-} + Ca(OH)_2 + 2H_2O \rightarrow Mg(OH)_2 + CaSO_4 \cdot 2H_2O$$
(brucite)   (gypsum)

$$xMg^{2+} + xSO_4^{2-} + xCaO \cdot SiO_2 aq. + 3xH_2O \rightarrow xCaSO_4 \cdot 2H_2O$$
(calcium silicate hydrate) $+ xMg(OH)_2 + SiO_2 aq.$

$$CaO \cdot SiO_2 \cdot aq. + 2Mg^{2+} + SO_4^{2-} + 2OH^- + 2H_2O \rightarrow MgO \cdot SiO_2 \cdot aq. + CaSO_4 \cdot 2H_2O$$
(Mg-silicate hydrate)

$+ Mg(OH)_2$

Sulfuric acid attack

$$Ca(OH)_2 + H_2SO_4 \rightarrow CaSO_4 \cdot 2H_2O$$
$$xCaO \cdot SiO_2 aq. + xH_2SO_4 + xH_2O \rightarrow xCaSO_4 \cdot 2H_2O + SiO_2 aq.$$

---

[1] Simplified. In some cases the accepted cement nomenclature is used (see Table 1).

Table 5: Continued

Ammonium sulfate attack
$$Ca(OH)_2 + (NH_4)_2SO_4 + 2H2O \rightarrow CaSO_4 \cdot 2H_2O + 2NH_4OH$$
$$CaO \cdot SiO_2 \text{ aq.} + (NH_4)_2SO_4 + 3H_2O \rightarrow SiO_2 \text{ aq.} + CaSO_4 \cdot 2H_2O$$
$$+2NH_4OH$$

Seawater attack
$$Mg^{2+} + Ca(OH)_2 \rightarrow Mg(OH)_2 + Ca^{2+}$$
$$CO_2 + Ca(OH)_2 \rightarrow CaCO_3 + H_2O$$
$$CO_2 + CaO \cdot SiO_2 \text{ aq.} \rightarrow CaCO_3 + SiO_2 \text{ aq.}$$
$$CaO \cdot SiO_2 \text{ aq.} + Mg^{2+} + SO_4^{2-} + Ca^{2+} + 2OH^- \rightarrow MgO \cdot SiO_2 \text{ aq.}$$
$$+CaSO_4 \cdot 2H_2O + Ca^{2+} + 2OH^-$$

C_3A or other Al-bearing phases. To decrease the possibility of this happening, ASTM recognizes two types of so-called *sulfate resisting* cements: Type II – moderate sulfate resistance and Type V – high sulfate resistance, both having decreased amounts of C_3A (Fig. 14). Although such cements do produce concrete with higher sulfate durability, use of sulfate-resisting cement is not, by itself, an adequate protection against sulfate attack. First of all, these cements are designed to protect concrete only against ettringite-type sulfate attack and are ineffective against the gypsum-type of attack or against sulfate salt crystallization. Second, such cements are ineffective in cases where Mg-sulfate is present in the aggressive medium. As shown in Table 5, the presence of magnesium leads to destruction of C-S-H (the "heart of concrete") in addition to the classical ettringite-type damage described above. The limited importance of cement type in the prevention of sulfate attack was recently discussed by Stark [15]. Lastly, as well publicized in the open literature, the use of sulfate resisting cements is only a secondary protection, the primary protection being impermeable concrete.

In concrete structures exposed to sodium sulfate solutions where some surfaces are exposed to evaporation due to temperature and humidity changes (e.g. partially exposed foundations, slabs on ground), the primary damage mechanism may be expansion due to repeated recrystallization of mirabilite ($Na_2SO_4 \cdot 10H_2O$) to/from thenardite (Na_2SO_4). (At temperatures below 32.4°C, mirabilite is formed; above this temperature, it recrystallizes.) This form of external attack is often incorrectly referred to as *physical sulfate attack* or as *sulfate salt crystallization distress*. It is a temperature-dependent process that leads to a repeated increase in volume and, thus, if the process happens at the concrete surface, can result in surface scaling. If the volume change happens deeper within the concrete matrix, it will lead to fatigue of the cement paste and a subsequent loss of cohesion. The repeated expansion-contraction is the origin of the term *physical* in describing this form of sulfate attack [9].

As mentioned above, so-called *physical sulfate attack* is often separated from other forms of *chemical* sulfate attack. This is incorrect for at least two reasons.

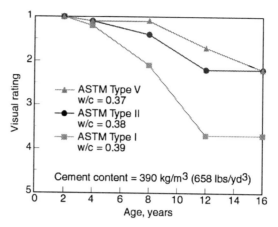

Figure 14: Performance of concretes made with different cements in sulfate soil. Type II and Type V cements have lower C_3A contents; this may improve sulfate resistance under some, though not all, soil conditions (Courtesy: Portland Cement Association).

First, like all crystallization processes, repeated recrystallization of mirabilite to/from thenardite is a *chemical* process with *physical* consequences; it is not unlike crystallization of calcium hydroxide or ettringite or any other salt. Second, while transported through the bulk concrete to the concrete evaporative surface, the ions in the sodium sulfate-rich pore solution may react with cement paste components to form the *classical* sulfate attack products (i.e. ettringite or gypsum or both) if the microenvironmental conditions are favorable. Chemical and microstructural changes, and thus the resulting damage, will occur.

Salt crystallization distress may also occur with salts other than sodium sulfate; $CaCO_3$ (calcium carbonate), $8.5CaSiO_3 \cdot 9.5CaCO_3 \cdot CaSO_4 \cdot 15H_2O$ (birunite), $NaHCO_3$ (nahcolite), $NaCl$ (halite), and $Na_3H(CO_3)_2 \cdot 2H_2O$ (trona) are typical examples.

Another form of external sulfate attack is referred to as *thaumasite sulfate attack*. Thaumasite, $Ca_3[Si(OH)_6] \cdot CO_3 \cdot SO_4 \cdot 12H_2O$, forms when concrete is exposed to both carbonate and sulfate ions, preferably at relatively low temperatures [16]. It typically leads to expansion of the matrix (ettringite formation) followed by loss of cohesion due to decomposition of C-S-H.

A special form of sulfate attack is the so-called *bacteriogenic attack*. This form of acid attack often occurs in sewage systems, where desulfovibrio bacteria transform organic sulfur to H_2S, which is subsequently oxidized to sulfuric acid:

$$\text{Organic sulfur} + (\text{desulfovibrio bacteria}) \rightarrow H_2S \tag{1}$$

$$H_2S + 1/2O_2 \rightarrow H_2O + S \tag{2}$$

$$H_2S + 3/2O_2 \rightarrow H_2O + SO_2 \tag{3}$$

then
$$SO_2 + H_2O + 1/2O_2 \rightarrow H_2SO_4 \tag{4}$$

Under other conditions:

$$S + H_2O + 3/2O_2 + \text{(thiobacilli bacteria)} \rightarrow H_2SO_4 \tag{5}$$

Over time, H_2SO_4 reacts with hydrated or anhydrous cement phases to form gypsum and various residual compounds.

The American Concrete Institute (ACI) Guide to Durable Concrete [5] recommends specific criteria for concrete designed for different sulfate environments (see Table 6). It is clear from this table that there are two important restrictions that apply simultaneously: (a) maximum water-to-cement ratio and (b) type of cementitious material. Thus, even if the structural design calls for relatively low strength, the durability criteria calling for both *low w/c and special cementitious systems* must be obeyed if desired durability is to be achieved, even if this may mean higher than needed strength. As discussed earlier, strength is not a measure of durability.

In a similar way, the British Standards have complex requirements covering cement type and level of sulfate or magnesium exposure, and recommendations calling for a combination of minimum cement content, maximum water-to-cement ratio, and cement type to be used (see Table 7) [17].

2.1.2 Prevention of sulfate attack

Because sulfate attack on concrete is not characterized by a single chemical process, when considering the best preventive measures one has to take into consideration the processes involved in concrete production and the environmental conditions to which the concrete will be exposed throughout its service life. However, some basic rules apply in most cases. The most important among them are:

1.   Minimizing access of water-soluble sulfates to the structure (e.g. proper structural design, drainage, protective barriers/coatings),

Table 6: ACI-recommended requirements for protection of concrete against external sulfate attack [5].

Severity of potential exposure	Water-soluble sulfate (SO$_4$) (%)	Sulfate (SO$_4$) in water (ppm)	w/c by mass, maximum	Cementitious materials requirement
Class 0	Less than 0.10	0–150	No requirement	No requirement
Class 1	0.10–0.20	150–1500	0.50	ASTM Type II or equivalent
Class 2	0.20–2.0	1500–10,000	0.45	ASTM Type V or equivalent
Class 3	2.0 or greater	10,000 or greater	0.40	ASTM Type V plus pozzolan or slag

2. Production of low-permeability concrete (e.g. use of low w/c mixture proportions, adequate cement content, proper consolidation, adequate curing), and
3. Intelligent selection of cementing materials (e.g. use of low-alumina, low-calcium cements, including sulfate-resisting cements and cements containing pozzolans or slag).

For reliable sulfate resistance, all the above preventive measures should be applied. Under special circumstances, such as heat-cured concrete or concretes exposed to severe environmental conditions (industrial, agricultural), additional preventive measures should be considered [5, 9, 18].

2.2 Industrial and agricultural chemicals

Concrete is seldom attacked by solid industrial or agricultural chemical waste, but may be easily damaged by certain kinds of liquid or water-soluble chemicals. If there is a pressure and/or temperature difference between the exposed and nonexposed concrete surfaces, the rate of damage may increase. Among the most damaging chemicals are various inorganic and organic acids (e.g. hydrochloric, nitric, phosphoric, sulfuric, acetic, formic, lactic acids), some alkaline solutions (e.g. concentrated sodium hydroxide, ammonium hydroxide), solutions of inorganic salts (e.g. aluminum-, ammonium- and magnesium- chlorides; sodium cyanide; ammonium nitrate; and ammonium-, calcium-, magnesium-, and sodium-sulfates), and other chemicals like liquid ammonia or gaseous bromine and chlorine. The mechanisms of action vary as the above cation-anion combinations may lead to acid attack, sulfate attack, decomposition of C-S-H, or other deleterious reactions and their combinations.

For additional information, see [19–22].

2.3 Atmospheric deterioration

2.3.1 Carbonation

Because Portland cement-based concrete always contains high proportions of calcium hydroxide and other calcium-containing compounds, exposure to carbon dioxide (CO_2) in the presence of moisture will induce formation of calcium carbonate. This may lead to several possible problems, especially, but not exclusively, in thin-wall products:

1. Shrinkage of the cement paste leading to cracking,
2. Decreased alkalinity (pH) of the cement paste enabling increased rate of corrosion of the reinforcing steel, and
3. Decreased wear resistance of carbonated surfaces.

Sources of aggressive carbon dioxide include groundwater or CO_2 in the air. Whereas carbonation of a concrete matrix by dry atmospheric CO_2 is a relatively slow process, in the presence of moisture or when dissolved in ground or rainwater, carbonation may be relatively rapid. Carbonation, like other chemical processes of

Table 7: UK recommendations for concrete exposed to sulfate attack (based on [17]: BS 5328; Part 1: 1997: Table 7; BRE Digest 363, 1991, Table 1).

	Exposure conditions: concentration of sulfate and magnesia					Recommendations		
Sulfate Class	SO_3 in ground water (g/l)	Mg in ground water (g/l)	SO_3 in soil or fill by acid extraction (%)	SO_3 in soil or fill by 2:1 water/soil extract (g/l)	Mg in soil or fill (g/l)	Cement group[b]	Minimum cement content (kg/m^3)	Maximum free water-to-cement ratio
1	<0.4	–	<0.24	<1.2	–	1, 2, 3	–	–
2	0.4–1.4	–	a	1.2–2.3	–	1	330	0.50
						2	300	0.55
						3	280	0.55
3	1.5–3.0	–	a	2.4–3.7	–	2	340	0.50
						3	320	0.50
4A	3.1–6.0	≤1.0	a	3.8–6.7	≤1.2	2	380	0.45
						3	360	0.45
4B	3.1–6.0	>1.0	a	3.8–6.7	>1.2	3	360	0.45
5A	>6.0	≤1.0	a	>6.7	≤1.2	As for class 4A plus surface protection	As for class 4A plus surface protection	As for class 4A plus surface protection
5B	>6.0	>1.0	a	>6.7	>1.2	As for class 4B plus surface protection	As for class 4B plus surface protection	As for class 4B plus surface protection

Note: Classification on the basis of groundwater samples is preferred. The limit on water-soluble magnesium does not apply for brackish water (having chloride content between 12 and 18 g/l).
aClassify on the basis of a 2:1 water/soil extract.
bRefers to cement group per (e.g. 3 refers to sulfate-resisting cement per BS 4027).

concrete deterioration, depends on concentration, relative humidity and temperature, and on the quality (primarily permeability) of the concrete. Industrial areas with high CO_2 content in the air may exhibit higher rates of concrete carbonation. The rate of atmospheric carbonation will be highest at relative humidities of about 50–75%.

The primary structural damage mechanism related to carbonation is corrosion of steel reinforcement. This aspect of concrete deterioration is discussed in more details in another part of this book.

2.3.2 Acid rain

The concrete matrix is a highly alkaline material. Thus, its resistance to acids is relatively low. However, for well-produced concrete, where the matrix permeability is low, concrete products and structures withstand occasional deterioration by atmospheric acids quite well.

The source of most atmospheric acid is sulfurous combustion gases that react with moisture to form low-concentration sulfuric acid. Other sources of acids include industrial or municipal sewage, as mentioned above.

2.3.3 Other deterioration mechanisms

The effect of seawater on concrete could have been categorized under sulfate attack or corrosion of reinforcement or otherwise but, because of the complex chemistry of seawater and the multiple potential deterioration mechanisms involved, it will be briefly discussed as a separate issue.

2.3.3.1 Exposure of concrete to seawater   Some of the most important and environmentally vulnerable concrete structures are located in marine environments. This enhanced vulnerability is related to continuous changes in temperature (including wetting/drying and freezing/thawing cycles), pressure changes in the tidal zone, reoccurring capillary suction, surface evaporation, and multiple chemical interactions with seawater components. Affected are the cement paste components, paste–aggregate interfaces, and the reinforcing steel. Deterioration of concrete exposed to seawater is known to be caused primarily by two independent chemical mechanisms: (a) corrosion of reinforcing steel embedded in concrete, and (b) deterioration of the concrete matrix itself. The rate of damage varies widely, depending upon the concrete quality, its maintenance, and the severity of the environment.

Damage to concrete by seawater is a result of the simultaneous action of multiple ions present in the water in different concentrations. The primary anionic components present are chlorides, sulfates, and carbonates, balanced primarily by cations Na^+, Mg^{2+}, and smaller amounts of K^+, Ca^{2+}. The typical concentration of SO_4^{2-} in seawater is about 2700 mg/l; that of Cl^- is about 20,000 mg/l.

The initial products of seawater reacting with concrete are aragonite, $CaCO_3$, and brucite, $Mg(OH)_2$, formed by the action of dissolved CO_2 and Mg^{2+}. Carbonic acid attack, enabled by the presence of small amounts of dissolved CO_2 in seawater, is based on carbonation of free calcium hydroxide, $Ca(OH)_2$, to form purely soluble *aragonite* rather than the usual calcitic form of calcium carbonate. The formation of metastable aragonite, in place of calcite, is promoted by the presence of Mg^{2+} ions. This is considered beneficial if the carbonation occurs only at the concrete surface, without reaching the $Ca(OH)_2$ within the concrete, because it may decrease the permeability of the concrete surface. Under different conditions, the $Ca(OH)_2$ can dissolve, leading to increased permeability.

$$CO_2 + Ca(OH)_2 \rightarrow CaCO_3$$
$$xCO_2 + xCaO{\cdot}SiO_2 \text{ aq.} \rightarrow xCaCO_3 + SiO_2 \text{ aq.}$$

(aragonite)

$Ca(OH)_2$ can also react with magnesium cations (ca 1,400 mg per liter) to form *brucite*, $Mg(OH)_2$, a highly insoluble mineral which, together with aragonite, is believed to densify the concrete surface, thus preventing or minimizing the ingress of seawater into the concrete structure. Such a reaction would induce very severe damage in permeable concrete, but is only minimal in well designed concrete.

$$Mg^{2+} + Ca(OH)_2 \rightarrow Mg(OH)_2 + Ca^{2+}$$

(brucite)

The products of the above reactions are believed to produce a surface skin consisting of a layer of brucite about $30\,\mu m$ thick overlaid by a layer of aragonite. This skin may protect well-produced concrete from further attack in permanently submerged regions or may, at least, slow down the progress of further corrosion. Additional phases formed in seawater attack may include magnesium silicates, gypsum, ettringite, and calcite, and also thaumasite in concrete mixes made with a carbonate-based aggregate.

It needs to be noted that these reactions are complex because they vary dramatically, depending upon the local environmental conditions (concentration and relative availability of ions in the water; chemical and microstructural properties of the cement paste in the concrete matrix; temperature, etc.). For example, there is a chemical 'competition' between the Mg^{2+}, sulfate and carbonate ions (capable of reacting with the Ca^{2+} originating from $Ca(OH)_2$ and C-S-H). Additionally, the Cl^- ions can not only suppress the presence of OH^- ions, thus changing the pH of the pore solution but, because of this pH change, can also influence the availability or reactivity of paste components, for example C_3A, thus enabling the potential formation of *Friedel's salt* (calcium chloro-aluminate hydrate, $C_3A \cdot CaCl_2 \cdot 10H_2O$) rather than ettringite.

The formation of ettringite in seawater attack typically does not lead to expansion and cracking of the concrete. It is believed that the formation of this phase is nonexpanding in the presence of excessive amounts of chloride ions. The primary deleterious effects result from the degradation of the C-S-H phase and its ultimate conversion to magnesium silicate. Sulfate attack may lead to gypsum formation. These conversions are ultimately responsible for the softening of the cement paste matrix and a gradual loss of strength.

$$xCaO \cdot SiO_2 \, aq. + 4Mg^{2+} + 4SO_4^{2-} + (4-x)Ca(OH)_2 \cdot nH_2O \rightarrow$$
$$4MgO \cdot SiO_2 \cdot 8H_2O + 4CaSO_4 \cdot 2H_2O$$

(magnesium silicate hydrate)   (gypsum)

In and above the tidal zone, a repeated cycle of water uptake and evaporation may result in salt crystallization, thus enhancing the deterioration of the concrete taking place. Still additional damage may by caused by mechanical erosion due to waves, solid debris or ice. In steel reinforced concrete, accelerated corrosion of the reinforcement can be induced by chloride ions that migrate into the concrete. Some of these issues, for example salt crystallization and steel corrosion, are discussed in other parts of this book [9, 23–26].

In conclusion, it should be emphasized again that the primary reason for deterioration of concrete by any one or all chemical mechanisms is access of water. Penetration of water or water-borne compounds (chlorides, sulfates, carbonates, etc.) into, within, and through concrete, and the resulting saturation and chemical changes of the concrete pore solution, lead to all critical durability problems we face in engineering practice: alkali–silica reaction, corrosion of reinforcement, freezing and thawing, and chemical attack – including internal and external sulfate attack. There are no exceptions! Thus, keeping unnecessary water out of concrete products and structures should be the primary goal of specifiers, structural designers, concrete producers, owners, and constructors/builders.

This can be done by maintaining a low w/c ratio, by effectively using chemical admixtures (water reducers, superplasticizers) and supplementary cementing materials (fly ash, slag, calcined clays, silica fume), and by proper processing (primarily curing and crack prevention). Crack prevention is a primary requirement: cracked concrete made with high quality aggregate and impermeable paste will *not* be durable.

3 Deterioration of aggregates in concrete

Aggregates make up about 70–80% of the volume of typical concrete. Thus, their quality and sensitivity to the environment of concrete exposure is of utmost importance. Depending upon the chemical and mineralogical composition, and on the geological origin of the rock used, aggregates may be sensitive to the microenvironment within the concrete matrix or to the external environment in which the concrete is used. Typical optical micrographs of sound concrete are presented in Fig. 15.

The most prevalent damage mechanisms affecting the aggregate portion of a concrete composite are alkali–aggregate reactions (caused by the reaction of alkalis with alkali-reactive aggregate particles) and repeated freezing and thawing. Whereas the former damage mechanism is controlled by environmental conditions within the concrete, the latter relates to repeated changes in temperature and humidity under ambient conditions.

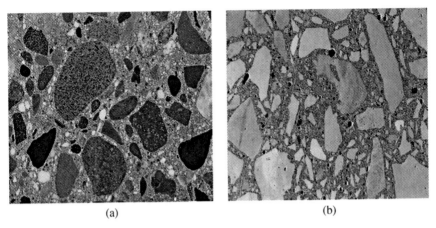

(a)   (b)

Figure 15: Cross-section of concrete made with (a) rounded siliceous gravel and (b) crushed limestone aggregate. In well-designed and processed concrete, cement paste completely coats each aggregate particle surface and fills all spaces between aggregate particles (Courtesy: Portland Cement Association).

3.1 Alkali–aggregate reaction

Alkali–aggregate reaction (AAR) is the generic name for interactions of aggregate components with alkali cations and OH^- ions present in the pore solution of concrete. The most important mechanisms are *alkali–silica reaction* (ASR), *alkali–carbonate reaction* (ACR), and various, less common, oxidation and hydration reactions of aggregate components (e.g. oxidation of pyrite, FeS_2; hydration of MgO).

Alkali–carbonate reaction can, under some circumstances, cause expansion and cracking. This is observed primarily with certain argillaceous (containing clay minerals) dolomitic limestones exhibiting specific microstructural features. The mechanism of damage is complex and somewhat ambiguous, and several theories were proposed to explain the damage. When dedolomitization of natural rock occurs, brucite is formed and alkalis present in the system are released. This, combined with the access of moisture and the presence of clay minerals in the system, may lead to swelling and subsequent damage to the concrete.

Oxidation of pyrite:

$$2FeS_2 + 7O_2 + 2H_2O \rightarrow 2FeSO_4 + 2H_2SO_4$$

(pyrite)

or

$$4FeS_2 + 15O_2 + 2H_2O \rightarrow 2Fe_2(SO_4)_3 + 2H_2SO_4$$

Hydration of magnesium oxide:

$$MgO + H_2O \rightarrow Mg(OH)_2$$

(brucite)

All of the above mechanisms depend on a supply of moisture and lead to expansion of the concrete followed by cracking and decreased durability.

Because these reactions – with the exception of water access – are primarily caused by internal rather than external environmental effects, and because – with the exception of ASR – they are not very common, we will briefly highlight only the problems related to alkali–silica interactions.

3.1.1 Alkali–silica reaction

Alkali–silica reaction is a potentially severe damage mechanism affecting concrete worldwide. The rate of reaction and the severity of damage vary widely, as they depend on numerous factors, including the type of aggregate (type and concentration of silica, alkali content), alkali content of the cement, concrete quality, structural design (access of water), and local reaction conditions such as temperature, humidity, and concrete permeability.

The most alkali-reactive mineral components of aggregates are opal, chalcedony, strained microcrystalline quartz, cristobalite, tridymite, and various natural and synthetic glasses (high surface area, devitrified, strained, etc.). These minerals may be present in the following aggregate types: cherts, limestones and dolomites, shales, schists, argillites, graywackes, quarzites, etc. The active component of these minerals is reactive, amorphous SiO_2, its reactivity induced by a disordered internal structure. In contrast to rocks containing crystalline silica, rocks containing amorphous reactive silica are usually porous and permeable. Thus, when hydroxyl ions become available, the siloxane (-Si-O-Si-) cross-linking is eliminated and the silica has the opportunity to hydrate and swell.

The primary factors affecting the deleterious chemical processes are presence of reactive, structurally disordered silica in the aggregate, alkalinity (concentration of OH^-) of the pore solution, and sufficient moisture. The active silica reacts with the alkali hydroxide present in the pore solution to form a hydrous alkali–silica–calcium gel of varying ratio of calcium to silica to sodium to potassium. This gel is osmotically active and, during swelling, can generate high pressures leading to expansion and cracking of the aggregate particles and the surrounding paste. Because the gel's viscosity depends on its composition, calcium must be a constituent of the gel to cause physical damage. Gel formed in the absence of calcium is fluid and thus cannot generate stresses. The gel originates inside the concrete but later, when cracking reaches the surface, fluid alkali–silica gel low in calcium is often extruded to the surface.

The mechanical damage caused by alkali–silica reaction includes cracking (expansion, map cracking), loss of strength, displacement, etc. An example of internal damage to aggregate in concrete is shown in Fig. 16. Typical cracking caused by

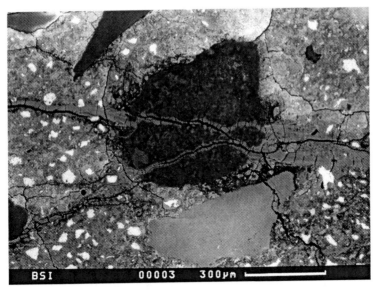

Figure 16: Cracking caused by ASR-induced expansion of aggregate. Backscattered SEM image (Courtesy: M. Thomas).

the alkali–silica reaction to a structure or product, so-called 'map cracking' or 'pattern cracking', is illustrated in Fig. 8.

As can be expected on the basis of the possible chemical interactions, the extent and progress of alkali–silica reaction can be strongly influenced and controlled by pozzolans (natural or artificial high-silica materials that are often used as supplementary materials in concrete; examples: high-silica fly ashes, microsilica, calcined shale). Their high amorphous silica content can reduce the hydroxyl ion concentration in the system, thus changing the reaction conditions needed for formation of the expansive gel. Replacing part of the cement with pozzolan, thus diluting the overall alkali content, typically does this. The positive effect of cement replacement is demonstrated in Fig. 17.

3.1.1.1 Prevention of alkali–silica damage   The primary preventive measures are as follows:

- Use of aggregate containing low concentrations of amorphous/reactive silica,
- Use of low-alkali cements, aggregates, and other cementing materials
- Protection of concrete from moisture, especially from water containing alkalis (e.g. seawater)
- Mixed design to produce cement paste of low porosity/permeability (low w/c, controlled cement content) and low free calcium hydroxide content (blended cements; high-silica supplementary cementing materials).

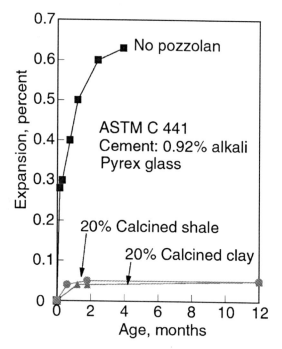

Figure 17: Reduction of alkali–silica reactivity by calcined clay and calcined shale (Courtesy: Portland Cement Association).

By proper selection of a combination of protective measures, erection of durable structural concrete is possible under most processing and environmental conditions.

3.2 Frost damage to aggregate: D-cracking

As is the case with other solid materials, the resistance of aggregates used in concreting to frost depends on their porosity and permeability. These are controlled by the overall pore structure (size distribution, shape, connectivity, etc.). Depending upon the size and accessibility of the pores, under humid conditions some of the pores will be water-filled; under saturated conditions, all the pores may be water-filled. Under freezing conditions, water in some pore sizes may freeze and, in near-saturated (critical saturation) conditions, the hydraulic pressures caused by ice formation may not be accommodated, thus leading to expansion.

Aggregate expansion near the concrete surface usually leads to 'pop-outs', wherein the portion of paste immediately between the aggregate particle and the surface is forced outward by the developing pressure and eventually breaks away. This process is, among other things, aggregate size dependent. If a large proportion of the aggregate particles are involved, especially when the cement paste is also susceptible, disintegration of the concrete, starting at edges and surfaces, may result.

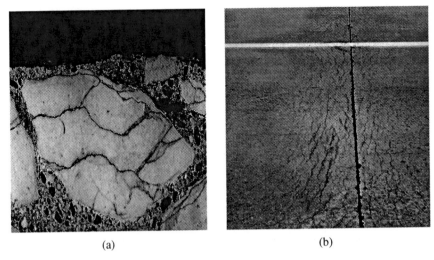

(a)   (b)

Figure 18: Deterioration of concrete due to freezing-thawing of aggregate. (a) D-cracking along a transverse joint caused by failure of carbonate coarse aggregate. (b) Fractured carbonate aggregate particle as a source of distress in D-cracking (magnification $2.5\times$) (Courtesy: Portland Cement Association).

A special form of aggregate damage is the so-called *D-cracking* of concrete pavements, caused by freeze–thaw deterioration of the aggregate within concrete. It resembles frost damage to concrete and is usually observable as closely spaced cracks at joints in concrete (see Fig.18a). This type of cracking, in addition to being dependent on the pore structure of the aggregate and the external environmental conditions (temperature and humidity changes), requires that the pavement base and sub-base be water saturated, thus enabling the aggregate in the pavement to become saturated and eventually to expand and crack (see Fig. 18b) [1]. Because the resulting damage is also aggregate size dependent, a reduction in maximum aggregate size and the provision of proper drainage are the best remedies against D-cracking.

4 Corrosion of steel reinforcement

Corrosion of the steel reinforcement in concrete structures is a significant problem throughout most of the world. Under ideal conditions, concrete is capable of chemically protecting the internal steel through the formation of a passivation film at the surface of the steel. Unfortunately, there are a number of conditions that can cause failure of the protective layer and allow corrosion to initiate, the primary factor being the introduction of chloride ions into the concrete. This can occur anywhere that reinforced concrete is exposed to seawater or deicing salts, making it a very widespread problem.

A complete description of the corrosion mechanism has been covered elsewhere in this work and will not be repeated here (see Chapter 6 by Andrade).

5 Damage to concrete by repeated freezing and thawing

If not properly designed and protected, concrete saturated with water can be susceptible to significant damage from exposure to freezing. When exposed to temperatures sufficiently low enough to induce freezing of the pore solution within the hydrated cement paste, internal tensile stresses develop that can grow large enough to induce fracture of the hardened paste. This type of deterioration is actually a combination of a number of mechanisms.

5.1 Mechanisms of freeze–thaw damage

The cumulative deterioration of saturated concrete under successive cycles of freezing and thawing is due to a number of concurrent physical and chemical mechanisms taking place during the freezing portion of the cycle. The damage created during this phase, in the form of microcracking within the hydrated cement paste, remains when the concrete thaws. During the successive freezing phase, the existing microcracks exacerbate the degradation. They allow more water to enter the paste and concentrate this excess water within the cracks, where the mechanisms associated with cumulative damage can cause further propagation of the cracks.

5.1.1 Hydraulic pressure

Unlike most materials, water actually expands upon freezing. This increase in volume is typically about 9%. Any time that water is frozen in a constrained space (i.e. within the hydrated cement paste pore structure) this expansion tends to result in the induction of tensile stresses within the surrounding material. Though the stresses created during the physical expansion of water as it changes state do contribute to the deterioration process, they do not constitute the sole, or even the primary, factor.

Since the volume of water increases as it freezes, the pressure within a confined space must increase as well. Also, the freezing point of any liquid is directly dependent upon the pressure under which the change in state takes place. Due to the nature of ice crystal formation, this relationship is not geometrically constant when the volume of the occupied space is below a certain maximum value. Within such small confines, ice crystal formation tends to increase the pressure in the allotted space much faster, and to a much higher degree, than in larger spaces. The result is a suppression of the freezing point that is dependent upon the volume of space occupied by the water.

In concrete, there is a very wide range in void and pore sizes in which the saturating water resides. The lower end of this size range extends below the threshold mentioned above, resulting in a differential freezing point from one region within the paste to another. Thus, the water within the concrete will not all change state at the same instant. Freezing will begin within the larger pores and progress to successively smaller pores in order of decreasing size. As the pore water freezes and increases in volume, the remaining unfrozen water is expelled from the larger cavities and forced into the smaller pores of the surrounding paste. A hydraulic pressure is created by this enforced flow, which induces tensile stresses that lead to cracking and failure of the paste structure [27].

5.1.2 Osmotic pressure

The water present within the concrete pore system is not pure. Within a very short period after entering the concrete, it becomes contaminated with various soluble substances (i.e. alkalis, free lime, chlorides). The concentration of this solution is not constant throughout the concrete, as the impurities are not found in equal amounts at all locations and, in the case of chlorides, may be available only from an exterior source. Additionally, the freezing point of such a solution will generally decrease as the concentration increases. Combining these factors results in another source of differential freezing, this time from areas of low concentration to areas of higher concentration.

As the impure water freezes, the concentration of soluble substances in the unfrozen water immediately adjacent to the freezing site increases, eventually surpassing that of the surrounding water. The resultant gradient in concentration creates an osmotic pressure as water migrates from regions of lower concentration in the surrounding paste [27]. This pressure can reach levels sufficient to induce damage to the paste.

5.1.3 Supercooling effect

An additional effect caused by the differential freezing progression from the larger capillary pores to smaller gel pores is a supercooling of the water in the latter areas [28]. Since the frozen water in the larger capillaries is in a low-energy state and the supercooled water in the gel pores is in a high-energy state, thermodynamic equilibrium is upset. To counter this imbalance, the supercooled water migrates toward the freezing front, effectively adding to the osmotic pressure effect.

5.2 Surface scaling

The previously described mechanisms refer to the bulk concrete matrix and do not consider the effect of boundary conditions. In many concrete structures, especially those that have a flat horizontal surface (i.e. sidewalks and bridge decks), a special type of freeze–thaw damage occurs. Small flakes or scales of concrete are exfoliated from the surface over time, resulting in complete destruction of the surface layer and loss of its protective presence. This type of damage is typically called *surface scaling* or *salt scaling*. Figure 19 [1] illustrates the typical appearance of an advanced case of surface scaling. It is believed that such damage is induced by a pair of mechanisms taking place at the interface between the concrete and the freezing environment.

5.2.1 Litvan's model

Litvan hypothesized that a vapor pressure develops between the supercooled pore water and ice on the surface of the concrete [29]. The vapor pressure gradient

Figure 19: Typical de-icer salt scaling of a concrete surface (Courtesy: Portland Cement Association).

induces desiccation of the concrete as the pore water migrates toward the surface. If the concrete possesses a less permeable surface layer, as is common in finished or formed surfaces, this water movement can result in pressure buildup just below the restrictive layer and eventually lead to mechanical failure.

5.2.2 Bi-directional freezing

When a frozen concrete structure is exposed to a thawing source that only thaws the outer portion of the concrete, a subsequent drop in temperature will result in a bi-directional freezing effect. Essentially, two freezing fronts are created, one moving outward from the inner, still frozen concrete and the other migrating inward from the surface. When these two fronts meet, an ice lens forms that creates a planar zone of weakness parallel to, and typically close to, the outer surface. The resultant expansion of the ice lens acts to force the surface layer outward, resulting in surface scaling.

This type of bi-directional freezing can occur when frozen concrete is exposed to incident solar radiation during the warm daytime hours and then re-freezes once the sun's energy is no longer available. It can also happen when de-icing salts are applied to a frozen concrete surface. The heat generated by the change in state of water from solid to liquid can thaw a very thin layer of the concrete. The latter case is where the expression *salt scaling* originated, as it tends to be a very prevalent damage form found throughout cold regions where de-icing salts are employed.

5.3 Overall effect

Freeze–thaw damage to concrete is actually a combination of the mechanisms described above, which results in a gradual deterioration of the concrete as the damage builds up over time. Freeze–thaw damage is deterioration through attrition. The amount of damage caused by a single freeze–thaw cycle is typically very small, and usually undetectable. It is only when large numbers of repeated freezing and thawing cycles occur that significant deterioration ensues. Consequently, freeze–thaw damage is mostly prevalent in temperate environments where the mean winter temperature approaches the freezing point of the pore solution.

In general, the concrete will tend to freeze during the night when temperatures are lowest and then thaw during the day due to rising temperatures and incident solar radiation. Under such exposure conditions, concretes can be subject to as many as 60 to 70 freeze–thaw cycles in a single winter [30, 31]. This total will, of course, vary dramatically depending upon the specific environment and exposure conditions of the concrete.

One significant exception is concrete placed within the tidal zone in cold environments. In regions where the ambient air temperature is sufficiently cold to freeze the pore solution, the rising water will thaw the concrete during each tidal cycle. Essentially, this situation can result in two freeze–thaw cycles per day, instead of the typical maximum of one cycle when dependent solely upon daily temperature fluctuations. Typical annual totals of 120–150 cycles are not uncommon [32], with some areas estimated to be as high as 200–300 [33].

5.4 Prevention of freeze–thaw damage

The damage mechanisms described above involve the induction of stresses within the concrete as water is forced, or drawn, through the pore structure. The magnitudes of these stresses are a function of: the permeability of the material through which the unfrozen water must pass, the amount of water available for ice formation, the degree of saturation of the concrete, the rate at which the ice forms, and the maximum distance to a free surface that can act as relief mechanism. Logically, protecting concrete from freeze–thaw damage should rely upon counteracting one or more of these factors.

Unfortunately, such an approach can be counterproductive when considering the issue of permeability, which would appear to suggest that a higher permeability will improve resistance to freezing and thawing. Increased permeability can easily be achieved through an increase in w/c ratio. However, doing so would be ineffective since it would also result in decreased strength and an increased volume of large pores, which is where the most readily freezable water is found [34].

In actuality, if the w/c ratio is sufficiently low, the resulting increase in strength combined with decreases in pore volume and permeability will produce a much more durable concrete. There will be less water available for ice formation due to the lower porosity and saturation will be much more difficult to achieve in conditions not involving continuous exposure to water, due to the lowered permeability. However, the w/c range needed to produce such a concrete (somewhere around 0.24) is too low to be considered practical in the field.

Attempts to control the degree of saturation of the concrete have been made through the use of waterproofing membranes or building designs preventing the pooling of water on concrete surfaces. Though helpful, neither of these methods is a universal solution that can be applied to all concrete structures. Membranes can degrade or wear off and the maximum slope of a member is often limited by the end use of the structure.

Relieving the stress buildup through the inclusion of empty air voids in the paste structure has proven to be the most applicable method. This approach is called *air entrainment* and is easily achieved through the use of specially designed chemical admixtures that are added to the fresh concrete during mixing. These admixtures consist of surfactants that essentially create a system of air bubbles within the cement paste. The air bubbles remain stable throughout the mixing, placing and setting procedures, resulting in empty voids scattered throughout the microstructure of the hydrated cement paste.

The surface film that induced formation of the air bubbles also creates a lower permeability film around the bubble, which later helps prevent water from entering the bubble under normal pressures. Thus, the air-entrained concrete will not reach full saturation under typical exposure conditions. Once freezing begins, however, these empty air voids act as pressure relief mechanisms. As unfrozen water is forced through the paste, pressure builds until it bursts through the void walls and into the empty space beyond, thus relieving the building pressure.

An effective air-entraining admixture creates an array of air bubbles that are dispersed evenly throughout the concrete but still close enough together such that the maximum distance from any point in the concrete to an air void is short enough to provide stress relief before the building pressure exceeds the strength of the hydrated cement paste itself. The average distance from any given point to an air void is designated as the *spacing factor*. Figure 20 [28] illustrates the effect of the spacing factor on the durability of a concrete mix.

It should be noted that there is a trade-off involved in implementing air entrainment for the freeze–thaw protection of concrete. As the air content increases, so does the overall porosity of the hydrated cement paste, resulting in a decrease in bulk strength. Typically, a 1% increase in entrained air will reduce compressive strength by about 5%, though this general relationship will vary from one concrete mix to another. In addition there is an upper limit on the amount of air that can be entrained before the voids begin to become interconnected and are unable to remain dry. Beyond this limit, the entrained air begins to lose its effectiveness. The result is an optimum air content that will provide maximum resistance to freeze–thaw without an excessive drop in strength. The relationship between these two factors is illustrated in Fig. 21 [28].

6 Mechanical deterioration of concrete

Concrete is susceptible to physical damage due to progressive mass loss induced by contact with moving objects or materials. This type of damage is also an attrition process, and usually takes many cycles of repetitive contact or a continuous exposure to the damage source.

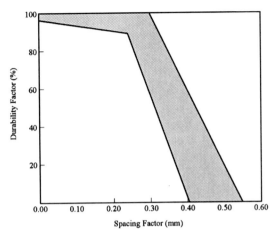

Figure 20: Relationship between air void spacing and durability of concrete [28].

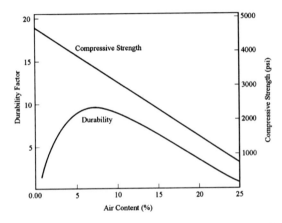

Figure 21: Relationship between air content, compressive strength, and durability of a concrete mix [28].

Mechanical deterioration typically affects the hydrated cement paste portion of the concrete and does very little damage to most aggregates since they are normally harder and more resistant. There are three primary types of mechanical deterioration: abrasion, erosion, and cavitation.

6.1 Abrasion

Abrasion damage is caused by direct physical contact with a moving object or material. Typically, this refers to wear induced by moving equipment, but can also apply to the flow of dry particles. Examples of the former would include vehicles or heavy equipment operating on pavements or floor slabs. The latter could refer to sand or aggregate stored in concrete bins and has even been observed in concrete silos used to store grain.

6.2 Erosion

Erosion refers to wear or deterioration that is induced by contact from solid particles suspended in a liquid. This type of damage is common in canal linings, spillways, sewage lines, and bridge piers. Anytime concrete is used to carry, direct, or control liquids containing significant quantities of silt, sand or other suspended particles, erosion will occur. Due to the continuous nature of such contact, erosion damage builds up slowly over time.

6.3 Cavitation

Cavitation is another source of damage common to concrete structures designed to carry liquids. The mechanism of damage in this case is the formation, and subsequent collapse, of vapor bubbles. Such bubbles form naturally within flowing water and later violently collapse when entering a turbulent or high velocity zone. Such regions are usually created by sudden changes in the direction of flow, possibly induced by a bend or change in geometry of a concrete conduit. Should these vapor bubbles be adjacent to the concrete surface while imploding, they will develop a suction force that attempts to pull the concrete away from the surface.

6.4 Prevention of mechanical damage

Protecting concrete from abrasion or erosion is typically done through the use of higher strength concrete. The American Concrete Institute Committee 201 recommends minimum values of 28 and 41 MPa, respectively, to provide adequate abrasion and erosion resistance for most exposures. Additionally, it is important to use hard, durable, wear-resistant aggregates and to minimize air contents relative to the exposure conditions.

Cavitation protection is harder to provide, since it is more a function of tensile strength than any other property. The tensile strength of concrete is much lower than its compressive strength and exhibits a much smaller range. It is also very difficult to achieve sufficient tensile strength to resist the forces generated by the cavitation mechanism. Instead, it is more effective to design water-carrying structures to eliminate the characteristics that cause the collapse of vapor bubbles. Abrupt changes in direction, slope, or geometry should be avoided.

7 Structural consequences of environmental damage

Depending on the type and extent of the damage, environmental degradation of concrete may lead to serious consequences in terms of structural failures, cost of repair, and indirect economic losses such as caused by delays in transportation and the inability of using the industrial and military infrastructures. Although catastrophic failures are uncommon, many concrete structures – especially those considered to be of lesser importance (e.g. residential, agricultural) – suffer environmental deterioration well before the expected need for maintenance. Atmospheric carbonation, inadequate freeze–thaw resistance, inadequately designed and produced concrete for high-sulfate environment, use of marginal aggregates all may cause serious degradation, especially when the quality of concrete processing is low, as is often the case. It can be stated indisputably that the most common reasons for structural failure of concrete products and facilities are not the materials used to make concrete, but the inappropriate design of the concrete itself (e.g. high permeability) and of the structure (such as access of water), as well the low quality of the workmanship.

Concrete remains to be an outstanding material of construction and will not be easily replaced in the future by other materials. Better appreciation of the in-depth relationship between its physicochemical and mechanical properties, combined with better utilization of the existing knowledge, may minimize structural failures to acceptable limits.

8 Economic considerations

The importance of concrete industry becomes obvious when one realizes that concrete is the most widely used man-made product and is second only to water as the most utilized substance by the humanity.

On the positive side, concrete and related industries employ hundreds of thousands of people worldwide, and the products produced by these industries are basic to industrial progress and societal well-being. Transportation and housing infrastructures, private and governmental construction, and even energy and water utilities all strongly depend on the productivity, efficiency, and quality of the concrete industry. For example, in the United States, the construction materials companies employ about 200,000 people and the gross product of the cement and concrete manufacturing segment of the construction industry exceeds $35 billion annually [35]. Yearly worldwide per capita production of concrete exceeds one metric ton.

On the negative side, among other challenges, the production of the basic concrete ingredient, namely Portland cement clinker, requires large amounts of fossil fuel energy and the by-products of the manufacturing process include such undesirable substances as CO_2 and high-alkali solid waste (e.g. kiln dust). The cost of energy and the cost of environmental compliance is very high. Also, if improperly produced or maintained, the concrete infrastructure repair and replacement are extremely costly.

It is for the abovementioned reasons that the environmental durability of concrete as a material and as structure is extremely important in economic terms and is being taken very seriously by the manufacturers and users of concrete. The main technical challenges of the concrete industry can be summarized as follows [36]:

- Process improvement throughout the lifetime of concrete, including design, manufacturing, transportation, construction, maintenance, and repair;
- product performance in terms of quality (durability, mechanical performance, etc.) and life-cycle cost and performance;

- energy efficiency in all aspects of concrete life cycle, especially in cement clinker manufacturing fuel consumption (this may require novel technologies and products);
- environmental performance, including use by the industry of recycled waste and byproducts from within and outside the concrete industry; and
- widespread transfer of existing technologies and timely identification, delivery, and implementation of newly generated research.

In the short term, durability of concrete infrastructure – thus its economic performance – can be improved by proper use of existing knowledge, adherence to standards and codes, and improved education of construction personnel and public.

References

[1] Kosmatka, S.H., Kerkhoff, B., & Perenchio, W.C., *Design and Control of Concrete Mixes*, 14th edn, Portland Cement Association, 2001.

[2] ASTM C 150-00, Standard Specification for Portland Cement, ASTM, 2000.

[3] Powers, T.C., Structure and physical properties of hardened Portland cement paste. *J. Amer. Ceram. Soc.*, **41**, pp. 1–6, 1958.

[4] Taylor, H.F.W., *Cement Chemistry*, 2nd edn, Thomas Telford Publishing: London, 1997.

[5] *Guide to Durable Concrete*, ACI 201-2R-01, American Concrete Institute, 2001.

[6] Hearn, N. & Young, F.J., W/c ratio and sulfate attack – a review. *Materials Science of Concrete Special Volume: Sulfate Attack Mechanisms*, The American Ceramic Society, pp. 189–205, 1998.

[7] Hooton, D.R., Thomas, M.D.A., Marchand, J., & Beaudoin, J.J. (eds.), *Materials Science of Concrete Special Volume: Ion Transport in Cement-Based Materials*, The American Ceramic Society, 2001.

[8] Skalny, J. & Marchand, J. (eds.), *Materials Science of Concrete Special Volume: Sulfate Attack Mechanisms*, The American Ceramic Society, 1998.

[9] Skalny, J., Marchand, J., & Odler, I., *Sulfate Attack in Concrete*, Spon Press: London and New York, 2002.

[10] Erlin, B. (ed.), *Ettringite – The Sometimes Host of Expansion*, ACI SP177, American Concrete Institute, 1999.

[11] Scrivener, K.L. & Skalny, J. (eds.), *Internal Sulfate Attack and Delayed Ettringite Formation: Proceedings of the International RILEM TC 186-ISA Workshop*, RILEM PRO 35, 2004.

[12] Thomas, M.D.A., Delayed ettringite formation in concrete: Recent developments and future directions. *Materials Science of Concrete VI*, eds. S. Mindess & J. Skalny, The American Ceramic Society, pp. 435–481, 2001.

[13] Diamond. S., Delayed ettringite formation – a current assessment. *Internal Sulfate Attack and Delayed Ettringite Formation: Proceedings of the*

International RILEM TC 186-ISA Workshop, RILEM PRO 35, eds. K.L. Scrivener & J. Skalny, pp. 178–194, 2004.

[14]   Skalny, J. & Marchand, J., Sulfate attack on concrete revisited. *Science of Cement and Concrete: Proceedings of the Kurdowski Symposium*, Krakow, June 2001, eds. W. Kurdowski, & M. Gawlicki, pp. 171–187, 2001.

[15]   Stark, D., Performance of concrete in sulfate environments. *R&D Bulletin*, RD129, Portland Cement Association, 2002.

[16]   British Research Establishment, *First International Conference on Thaumasite in Cementitious Materials: Book of Abstracts*, Garston, Watford, UK, June 19–21, 2002.

[17]   British Standard: BS 5328; Part 1: 1997: Table 7.

[18]   Hobbs, D.W., *Minimum Requirements for Durable Concrete*, British Cement Association, 1998.

[19]   Biczok, I., *Concrete Corrosion – Concrete Protection*, 8th edn, Akadémiai Kiadó: Budapest, 1972.

[20]   Portland Cement Association, Effect of Various Substances on Concrete and Protective Treatments, Where Required, Publication No. ISOO1T, Skokie, IL, 1968.

[21]   Reinhardt, H.W. (ed.), *Penetration and Permeability of Concrete: Barriers to Organic and Contaminating Liquids*, RILEM Report 16, E&FN Spon: London, 1997.

[22]   Bajza, A. Corrosion of hardened cement paste by $NH_4\ NO_3$ and acetic and formic acids. *Pore Structure and Permeability of Cementitious Materials*, Materials Research Society: Pittsburgh, Vol. 137, pp. 325–334, 1989.

[23]   Boyd, A.J., Mindess, S. & Skalny, J., Designing concrete for durability. *Materiales de Construccion*, **51(263–264)**, pp. 37–53, 2001.

[24]   Thomas, M.D.A., Bleszynski, R.F. & Scott, C.E., Sulfate attack in marine environment. *Materials Science of Concrete Special Volume: Sulfate Attack Mechanisms*, eds. J. Skalny & J. Marchand, The American Ceramic Society, pp. 301–313, 1999.

[25]   Mehta, P.K., *Concrete in Marine Environment*, Elsevier Applied Science: London, 1991.

[26]   Mehta, P.K., Sulfate Attack in marine environment. *Materials Science of Concrete Special Volume: Sulfate Attack Mechanisms*, eds. J. Skalny & J. Marchand, The American Ceramic Society, pp. 295–300, 1999.

[27]   Powers, T.C., The physical structure and engineering properties of concrete, Portland Cement Association, Bulletin 90, Skokie, IL, 1958.

[28]   Mindess, S., Young, J.F. & Darwin, D., *Concrete*, 2nd edn, Prentice-Hall: Upper Saddle River, NJ, 2003.

[29]   Litvan, G.G., Frost action in cement in the presence of de-icers. *Cement and Concrete Research*, **6**, pp. 351–356, 1976.

[30]   Sturrup, V., Hooton, R.D., Mukherjee, P. & Carmichael, T., Evaluation and prediction of concrete durability – Ontario Hydro's experience. *Katherine and Bryant Mather International Conference on Concrete Durability*, ACI, SP-100, Detroit, USA, 1987.

[31]   Boyd, A.J., Salt Scaling Resistance of Concrete Containing Slag and Fly Ash, MASc Thesis, University of Toronto, Toronto, Canada, 1995.

[32]   Zhang, M.H., Bremner, T.W. & Malhotra, V.M., Effect of the type of Portland cement on the performance of non-reinforced concrete in marine environment at Treat Island, Maine. *5th International Conference on Durability of Concrete*, CANMET/ACI, Barcelona, Spain, June 2000.

[33]   Gilbride, P., Morgan, D.R. & Bremner, T.W., Deterioration and rehabilitation of berth faces in tidal zones at the Port of Saint John. *2nd International Conference on Concrete in Marine Environment*, ACI, SP-109, Detroit, USA, 1988.

[34]   Mehta, P.K., *Concrete: Structure, Properties, and Materials*, Prentice-Hall: Englewood Cliffs, NJ, 1986.

[35]   US Department of Commerce, US Bureau of Census, Manufacturing Industry Series, 1997.

[36]   *Vision 2030: A Vision for the U.S. Concrete Industry*, American Concrete Institute, 2001.

CHAPTER 6

Corrosion of steel reinforcement

C. Andrade
Institute of Construction Science 'Eduardo Torroja', CSIC,
Madrid, Spain.

1 Principles of corrosion

Concrete is the most widely used construction material in the world. Many kinds of materials, elements and structures are fabricated with cement-based mixes. Reinforced concrete was industrially developed at the beginning of the 20th century, and it has stimulated tremendous developments in housing and infrastructures.

Reinforced and prestressed concrete represents a very successful combination of materials, not only from a mechanical point of view but also from a chemical perspective, because the hydrated cement is able to provide to the steel an excellent protection against corrosion. This chemical compatibility allows for the composite behaviour of reinforced concrete and is the basis of its high durability.

The composite action occurring in the steel–concrete bond may be unlimited in time while steel remains passive. The study of the conditions leading to reinforcement corrosion is then of high importance because corrosion may significantly affect the load-bearing capacity of reinforced or prestressed concrete.

The natural state of metals is their oxidized state. Metals can be found in nature in the form of oxides, carbonates, sulphates, etc. (minerals). In a pure state only the so-called 'noble' metals can persist in contact with the environment without undergoing oxidation. For the practical use of a metal, a certain energy is invested in its reduction from the natural mineral state. The metal then presents the tendency to liberate this energy to attain its lower energy state. The process by which a metal returns to its mineral state is known as 'corrosion'. Corrosion is therefore the process by which the metal passes from its metallic state at 'zero' valence to its oxidized state liberating electrons. For iron it can be simply written as:

$$Fe \rightarrow Fe^{2+} + 2e^-$$

The electrons are transferred to other substances (oxygen, carbonate, sulphate, etc.) in order to become a neutral substance.

The mechanisms for the transfer of electrons from the metal to another substance can basically occur in two manners: directly or through an aqueous solution. The former is produced at high temperatures where water cannot exist in liquid form, while *aqueous corrosion* is the most common mechanism and it develops at normal temperatures.

Corrosion in the presence of liquid water occurs by an electrochemical mechanism [1]. Chemical reactions and redox processes occur simultaneously. Thus, metallic zones with different electrical potential due to the metal being in contact with a heterogeneous (in concentration) electrolyte or due to heterogeneities in the metal itself are the driving force of chemical reactions involving the exchange of electrons. The process occurs as in a battery where the oxidation of the metal takes place in the anodic zone, as shown in the first reaction, and a reduction takes place in the cathodic zone. For the case of neutral and alkaline electrolytes, the most common cathodic reaction is:

$$O_2 + 4e^- + 2H_2O \rightarrow 4OH^-$$

Therefore the electrons liberated in the anode circulate through the metal to the cathodic zones where they are consumed inducing the reduction of a substance (e.g. oxygen in the second reaction). The final corrosion product would be $Fe(OH)_2$, $Fe(OH)_3$ or some oxyhydroxides derived from them.

For the case of acid solutions the most common cathodic reaction in the reduction of protons

$$2H^+ + 2e^- \rightarrow H_2 \uparrow \text{(gas)}$$

Figure 1 shows the development of the corrosion cell and illustrates the need to have a continuous 'circuit' for the corrosion to progress, as in the case of batteries. The elements of a corrosion cell in aqueous-type corrosion are:

1. the anode, where the metal is dissolved;
2. the cathode, where a substance is reduced and takes up the electrons liberated by the metal during its oxidation;
3. continuity across the metal between anode and cathode;
4. continuity across the electrolyte for the chemical substances to move and become neutral.

The lack of electrolyte or of metal connexion will stop the corrosion and the development of any anodic or cathodic zone.

1.1 Corrosion morphology

The corrosion may progress by a uniform dissolution of the whole surface (Fig. 2a) or by a local attack which, when it is very localized, is called 'pitting' corrosion (Fig. 2b). It may also progress at the microscopic level when it is called 'inter- or trans-granular' (Fig. 2c) attack as metal grains are very locally affected.

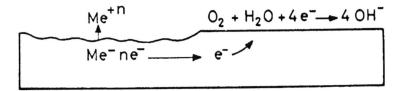

Figure 1: The metal dissolves in the anodic zones releasing metal ions and electrons. The latter are consumed in the cathodic zones reducing another substance such as oxygen.

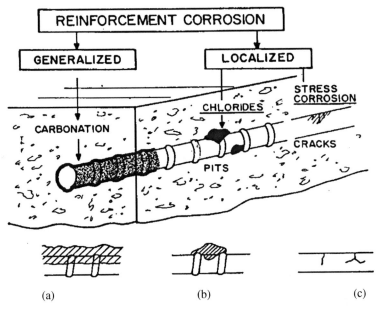

Figure 2: Types of corrosion of reinforcement: (a) carbonation, (b) chloride attack and (c) stress corrosion cracking.

1.2 Notions of electrochemical potential

Not all the metals have the same tendency to oxidize, that is, not all the metals are equally reactive. The activity of metals when in contact with an electrolyte can be expressed through Nernst equation:

$$E = E_0 + \frac{RT}{nF} \ln k, \tag{1}$$

where E is the actual potential, E_0 is the so-called 'standard potential', R is the gas constant, T is the absolute temperature, n is the number of electrons exchanged,

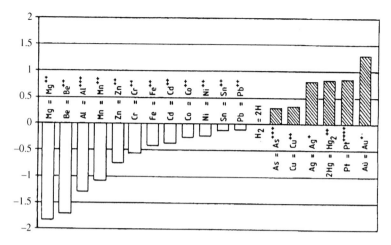

Figure 3: Nernst potentials shown in a graphic form. Positive values indicate noble metals and negative values indicate active ones.

F is the Faraday number and k is the equilibrium constant of the ions present in the electrolyte.

k takes different forms depending on the type of reaction. For the Fe(II)/Fe (metal) system, k is represented by the activity of the ion Fe^{2+} in the solution. Thus, if this activity is 10^{-3} mol/l, as $E_0 - 0.44$ V (SHE), the equilibrium potential E is [2]:

$$E_0 - 0.44 + \frac{0.059}{2}(-3) = 0.527 \text{ V.} \qquad (2)$$

The potential is a measure of the facility of exchanging of electrons across the metal/electrolyte interface and of the ease of the reduction reaction. The absolute potential values cannot be determined and therefore they are given by comparison with a redox reaction (e.g. that of hydrogen in the third reaction) which is taken as reference. That is why corrosion potentials are expressed with reference to the hydrogen electrode, or to any other reference electrode (calomel, copper/copper sulphate, etc.). These electrodes have very rapid redox exchanges and therefore constant reference potentials. Figure 3 shows the values of the most common reference electrodes taken as 'zero', that of hydrogen electrode.

1.3 Pourbaix diagrams

The potential exhibited by an electrode in a particular electrolyte depends on a set of factors: (a) the standard potential of the anodic and cathodic reactions, (b) the temperature, and (c) the composition of the electrolyte, that is, its ionic concentration. The last factor is very well expressed by Pourbaix [3] using pH–potential diagrams (Fig. 4). These diagrams represent the conditions of potential and pH where a particular corrosion reaction is thermodynamically favourable (Nernst equation is used

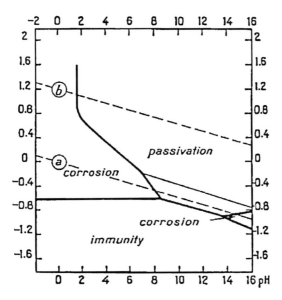

Figure 4: Pourbaix's diagram for Fe.

for the analysis of reactions). From the results the diagrams can be divided in three main zones: corrosion, passivity and immunity: The metals, when in contact with an aqueous solution may corrode, or become passive by the generation of a very stable oxide, or they may remain in a region of potentials where oxidation is not thermodynamically feasible (immunity).

Figure 4 shows Pourbaix's diagram for the Fe. The two lines a and b indicate the regions of water electrolysis. Below line b, water evolves to hydrogen gas and hydroxides: $2H_2O + 2e^- \rightarrow H_2 + 2OH^-$, and above line a, to oxygen and protons: $H_2O + 2e^- \rightarrow 2H^+ + O_2$.

1.4 Polarization

In spite of the usefulness of Pourbaix's diagrams, they represent only equilibrium conditions and therefore, kinetic aspects are not taken into consideration in them. Kinetic aspects have to be studied by following the changes in potential (overpotential) from equilibrium conditions (those indicated in Pourbaix's diagrams). Any change in potential from equilibrium conditions is due to the passage of a current and the phenomenon is called 'electrode polarization' [1, 4]. In general, it can be written as:

$$I = \frac{E_c - E_a}{R}, \tag{3}$$

where I is the current applied externally or recorded (if the change is induced in the potential), $E_c - E_a$ is the change in potential (overpotential or polarization) and R is the electrical resistance of the circuit.

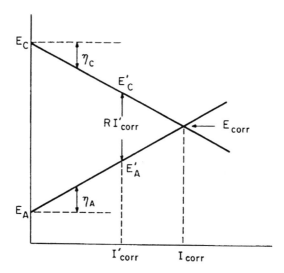

Figure 5: Evans diagram voltage/current for a corrosion cell in which the polarization curves have been linearized.

Although the relation between I and $(E_c - E_a)$ is generally not linear, Evans [1] introduced a simple representation which is used very much for the sake of illustration. Figure 5 depicts an Evans diagram. Evans diagrams helps to follow the evolution of the anodic, E_a, and cathodic, E_c, potentials with the passage of an external current, I. The slope of the lines represents the polarization, and their distance depends on the electrical resistance of the circuit, R.

A study of the types of polarization was carried out by Tafel in 1905 [4] to establish that for high enough polarization, the relation between the overpotential $\eta = E_c - E_a$ and the current is:

$$\eta = a + b \log i, \tag{4}$$

where a is the order at the origin and b is the slope. This expression enables the calculation of 'Tafel slopes', the basic parameter to study the kinetics of corrosion reactions, and the so-called 'polarization curves'.

1.5 Calculation of corrosion rate

It was in 1957 when Stern and Geary [5] in a study of the behaviour of polarization curves established that around the corrosion potential, i.e. when the current or overpotential is very small, the polarization curves could be linearized and therefore the slope of the curve (Fig. 6) is directly proportional to what was named 'polarization resistance', R_p:

$$\left(\frac{\Delta E}{\Delta I} \right)_{\Delta E \to 0} = R_p. \tag{5}$$

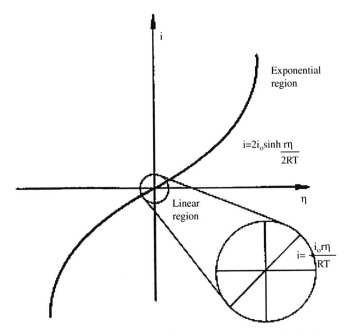

Figure 6: The polarization curve around the corrosion potential is linear and can be used to measure the polarization resistance, R_p. i = current density, i_0 = exchanged current, η = overpotential and $r = F$ = Faraday constant.

This resistance to polarization, determined by very small overpotentials, is inversely proportional to the corrosion current or rate through:

$$I_{corr} = B/R_p, \qquad (6)$$

where B is a constant which is based on Tafel slopes (b_a, b_c):

$$B = \frac{b_a b_c}{2.3(b_a + b_c)}. \qquad (7)$$

As the overpotentials needed to determine I_{corr} are very small, the results of the method are non-destructive. The technique has been much developed and is now the basis for the current determination of corrosion rates in many metal/electrolytes systems. It was applied to the study of reinforcement corrosion by Andrade and Gonzalez [6] in 1970 and has been implemented in corrosion-rate-meters for measuring in real-size structures by Feliú et al. [7].

2 Damage to structural concrete due to corrosion

The main factors leading to reinforcement corrosion are shown in Fig. 2: Carbonation induces a generalized corrosion while the presence of chloride ions in the surroundings of the steel provokes localized corrosion. Stress corrosion cracking (SCC) has been detected in prestressing and post-tensioned structures [8].

2.1 Carbonation

Carbon dioxide is present in the air in proportions ranging from 0.03% to 0.1% depending on the local contamination. This gas reacts with the alkalinity of cement phases to give $CaCO_3$ as the main reaction product. This process leads into a decrease of the pH value of the pore solution from pH > 13 to pH < 8. At this neutral pH value, the passive layer of steel disappears and a general and uniform corrosion starts.

The CO_2, being a gas, penetrates through pore network of concrete. If the pores are filled with water, the gas dissolves in the liquid water and the penetration is then very slow (Fig. 7). If the pores are not saturated with water, then the CO_2 can easily penetrate by diffusion and reach internal parts of the concrete cover.

The carbonation reaction is then produced between the CO_2 gas and the alkaline ions present in the pore solution. Thus, in schematic form:

$$CO_2 + H_2O \rightarrow CO_3^{2-} + 2H^+$$

$$CO_3H_2 + NaOH \text{ or } KOH \text{ or } Ca(OH)_2 \rightarrow Na_2CO_3 + H_2O$$
$$K_2CO_3 + H_2O$$
$$CaCO_3 + H_2O$$

LOW R.H.   HIGH R.H.   SATURATED

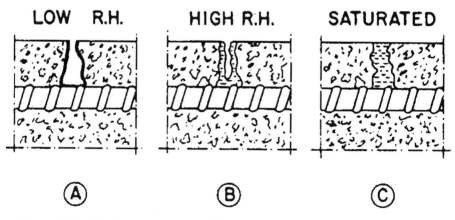

Figure 7: Simplified representation of the degree of saturation of concrete pores (RH, relative humidity). (A) CO_2 can easily penetrate, but the lack of moisture avoids carbonation. (B) Optimum moisture content for carbonation. (c) CO_2 penetrates very slowly when dissolving in water.

The neutralization of the pore solution, on the one hand, includes the leaching of Ca from the Ca-bearing cement phases, starting with C-S-H, and on the other, provokes the lowering of the pH of the pore solution.

Carbonation proceeds then in the water of the pore solution. If the concrete is very dry (Fig. 7), carbonation is not feasible due to the lack of water for the reaction to be produced. The maximum carbonation rate is noticed then at intermediate relative humidities (RHs) between 50% and 70% approximately. The oxides generated induce cracking of the cover through cracks parallel to the reinforcements (Fig. 8).

2.2 Chloride attack

The two main sources of chlorides for concrete structures are marine environments and the use of deicing salts in roads in cold climates. Chloride ions can be present in the concrete, either because they are introduced in the mixing materials, or because they penetrate from outside dissolved in the pore solution. While carbonation implies a gas that needs the pore network partially empty, chloride ions penetrate if the pores are filled with water, or a marine fog impregnates the concrete.

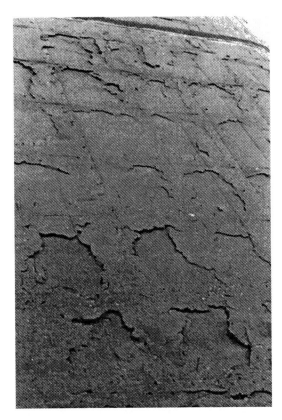

Figure 8: Crack pattern due to corrosion induced by carbonation.

Figure 9: Localized attack due to chlorides.

Chlorides provoke a local disruption of the passive layer leading to localized attack (Fig. 9). The disruption is due to a local acidification induced by the protons liberated from the formation of iron hydroxides [4, 9]:

$$Fe^{2+} + 2H_2O \rightarrow Fe(OH)_2 + 2H^+$$

Depending upon the extension of the corrosion, cover cracking can appear or not. In submerged structures, cracking is often not induced.

2.2.1 Chloride threshold

For the attack to develop it is necessary that a certain amount of chlorides reach the steel surface. This amount is known as the chloride threshold and it is not an unique quantity because it is influenced by many parameters. The main influencing factors are:

- cement type: fineness, C_3A of cement, SO_3 content;
- presence of blending agents and their composition;
- curing and compaction;
- moisture content in concrete pores;
- steel type and surface finishing;
- local oxygen availability.

Too many parameters have prevented the identification of a single threshold value in the past [10]. More recent work [11] has, however, helped to identify the electrical potential as the controlling factor of the chloride threshold. Figure 10 shows the dependency. The same concrete induces different corrosion potentials, and this explains why the same concrete exhibits different chloride thresholds. The different corrosion potentials appear because of differences in the parameters listed above.

Figure 10 indicates that there is range of corrosion potentials (from around $-200\,mV_{SCE}$ to more noble potentials) in which the chloride threshold shows a minimum. When the potential is in a more negative/cathodic region, the amount of chloride ions inducing corrosion increases in coherence with the principles of cathodic protection [12]. Below a certain potential, the metal enters into the region of immunity of Pourbaix's diagram.

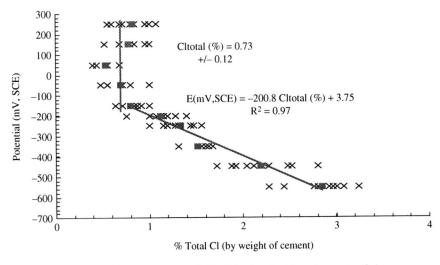

Figure 10: Dependence of chloride threshold on the potential.

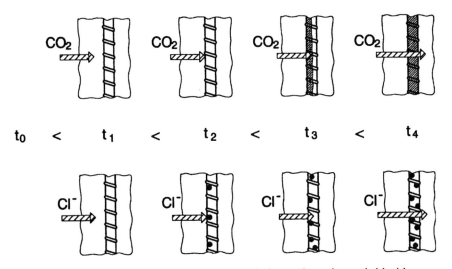

Figure 11: Progressive events of depassivation during carbonation and chloride penetration.

The usual chloride threshold specified in codes and standards is 0.4% by cement weight, not far from the minimum value obtained in the tests shown in Fig. 11, which was 0.7% by weight of cement.

The depassivation cannot be understood as an instantaneous process. Depassivation is a period of time during which the steel progressively becomes actively corroding (Fig. 11). Events of activity–passivity may develop during a long period without negligeable loss of steel cross-section being caused.

Figure 12: Cracks induced by a certain stress applied to a prestressing wire in bicarbonated solution.

2.2.2 Stress corrosion cracking

SCC is a particular case of localized corrosion. It occurs only in prestressing wires, when wires of high-yield point are stressed to a certain level and are in contact with a specific aggressive medium.

The process starts with the nucleation of microcracks at the surface of the steel (Fig. 12). One of the cracks may progress to a certain depth, resulting in a high crack velocity due to which the wire breaks in a brittle manner in relatively short time.

The mechanism of nucleation, and mainly of progression of SCC, is still subject to controversy. Nucleation can start in a surface non-homogeneity, spots of rust or inside a pit. The progression in enhanced by the generation of atomic hydrogen at the bottom of the crack. Of the several mechanisms proposed to explain the process, the one based on the concept of 'surface mobility' seems to best fit the experimental results. Surface mobility assumes that the progression of the crack is not of electrochemical nature but due to the mobility of atomic vacancies in the metal–electrolyte interface [13].

The only way to diagnose the occurrence of SCC is through the microscopic examination of the fractured surfaces in order to identify the brittle fracture. Thus, Fig. 13a shows a ductile fracture of a prestressing wire and Fig. 13b a brittle one (no striction is produced).

In the SCC phenomenon the metallographic nature and treatment of the steel plays a crucial role. Thus, quenched and tempered steels are very sensitive while the susceptibility of cold drawn steels is much lower. The use of the former is forbidden for prestressing in many countries.

2.3 Service life of reinforced concrete

When reinforced concrete started to be industrialized, it was believed that the material is going to have an unlimited durability. This is due to the supposition that cement alkalinity provides a chemical protection for the steel while the concrete cover is a physical barrier against contact with the atmosphere.

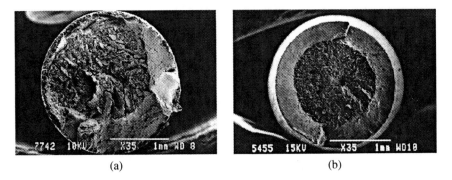

(a)   (b)

Figure 13: Fracture of a prestressing wire: (a) brittle; (b) ductile.

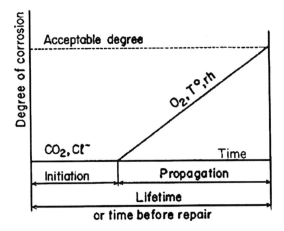

Figure 14: Service life model for reinforcement corrosion.

However, after 20–50 years of exposure to different environments, steel reinforcement exhibits corrosion, this being one of the most serious durability problems due to the economical consequences of repairs. Prediction of service life or time to corrosion has generated an increasing interest during the past two decades.

Judging the more or less theoretical proposals for service life prediction, such as that of ASTM E 632-81 until present situation, where concrete Codes and Standards try to incorporate durability chapters, it can be said that the most well-known service life model was the one published by Tuutti [14] in 1982. Figure 14 shows the model which specifies two periods for service life:

$$t_1 = t_i + t_p. \tag{8}$$

Service life in this case is defined as t_1. The initiation period, t_i, comprises the time taken by the aggressive agents (chlorides or carbonation front to reach the reinforcement or passivate the steel). This is the relevant period if depassivation is identified as the end of service life. However, if a certain amount of steel deterioration is considered as part of the design life, a period of propagation, t_p, can be considered until a certain 'unacceptable degree' of corrosion is reached.

This first proposal has been improved by other studies [15, 16]. The most common methodology followed at present when trying to design for reinforcement protection includes the following steps:

1. identification of aggressivity of the environment;
2. definition of the length, in years, of the service life, in addition to considering some special actions of maintenance or supplementary protection methods;
3. consideration of a calculation method for the attack progression;
4. specification of minimum concrete quality and cover depth.

2.3.1 Environmental aggressivity

Environmental actions are responsible for the lack of durability of reinforcement as they are the source of temperature cycles, of water supply through rain and snow, and of carbonation and chlorides. In general, no significant damage is noticed in indoor conditions, provided that no water is present due to leaking of pipes or lack of roof tightness.

Exposure classes are usually identified in Codes. Table 1 shows those specified in Eurocode 2 (CEN-1992 – part 1). In it, carbonation and chloride-containing ambients are separately considered.

2.3.2 Length of service life

In numerous standards, service life is defined as 'the period of time in which the structure maintains its design requirements of: safety, functionality and aesthetics without unexpected costs of maintenance'.

Services lives of 50–120 years are usually taken in relation to the types of structures. The explicit definition of length of service life has technical as well as legal implications, as civil responsibilities are involved for the designer and the constructor. Therefore, it should be considered as a reference period for the calculations, avoiding liabilities when these are not well specified.

2.3.3 Supplementary protection methods

In very aggressive environments or when the concrete element cannot have a sufficiently thick cover, supplementary protection methods should be specified from a design phase. The main methods are:

- cathodic protection,
- galvanized or stainless steel reinforcement,
- corrosion inhibitors,
- concrete coatings.

The main advantages and disadvantages of these methods are briefly outlined in Fig. 15 [17].

Table 1: Exposure classes in EN-206.

Class designation	Description of environment	Informative examples where exposure classes may occur
1. No risk of corrosion or attack		
X0	For concrete without reinforcement or embedded metal: all exposures except where there is freeze–thaw, abrasion or chemical attack	Concrete inside buildings with very low air humidity
	For concrete with reinforcement or embedded metal: very dry	
2. Corrosion induced by carbonation		
Where concrete containing reinforcement or other embedded metal is exposed to air moisture, the exposure shall be classified as follows:		
Note: The moisture condition relates to that in the concrete cover to reinforcement or other embedded metal but, in many cases, conditions in the concrete cover can be taken as reflecting that in the surrounding environment. In these cases, the classification of the surrounding environment may be adequate. This may not be the case if there is a barrier between the concrete and its environment.		
XC1	Dry or permanently wet	Concrete inside buildings with low air humidity
		Concrete permanently submerged in water
XC2	Wet, rarely dry	Concrete surfaces subject to long-term water contact
		Many foundations
XC3	Moderate humidity	Concrete inside buildings with moderate or high air humidity
		External concrete sheltered from rain
XC4	Cyclic wet and dry	Concrete surfaces subject to water contact, not within exposure class XC2
3. Corrosion induced by chlorides other than from sea water		
XD1	Moderate humidity	Concrete surfaces exposed to airborne chlorides
XD2	Wet, rarely dry	Swimming pools
		Concrete exposed to industrial waters containing chlorides
XD3	Cyclic wet and dry	Parts of bridges exposed to spray containing chlorides
		Pavements
		Car parks
4. Corrosion induced by chlorides from sea water		
XS1	Exposed to airborne salt but not in direct contact with sea water	Structures near to or on the coast
XS2	Permanently submerged	Parts of marine structures
XS3	Tidal, splash and spray zones	Parts of marine structures

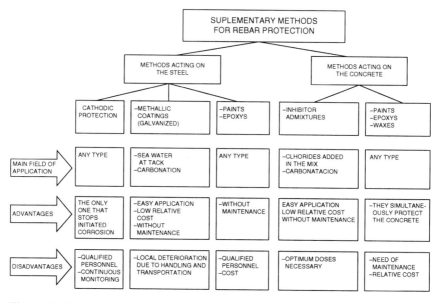

Figure 15: Supplementary methods of protection against reinforcement corrosion.

2.3.4 Calculation methods for attack penetration

The two main causes of corrosion, as previously described, are: the carbonation of concrete cover and the penetration of chloride ions.

Carbonation usually progresses by a diffusion mechanism while chlorides may penetrate also with a combination of absorption and diffusion (tidal or splash zones). Hydrostatic pressure may also lead into permeation. That is, among these three mechanisms, the most common are diffusion and absorption. They are known to follow the law of the 'square root of time':

$$x = K\sqrt{t}, \tag{9}$$

where x is the attack penetration depth (mm), t is the time (s) and K is a constant, depending on concrete and ambient characteristics. This square root law may be plotted on a log–log scale as shown in Fig. 16, and as suggested by Tuutti [14], which is very convenient and general.

2.3.4.1 Carbonation.

It is known to be a consequence of the CO_2 gas penetration through water empty pores and its dissolution in the pore water followed by the reaction with the alkaline compounds present there. Calcium carbonates precipitates and the pH of the pore solution drops to values near neutrality. At these neutral pH the steel depassivates and starts to corrode.

The *carbonation* rate has been modelled by many researchers, although only three models will be commented. Those are the models proposed by Tuutti [14],

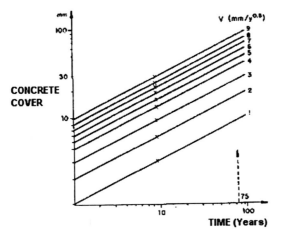

Figure 16: Representation of the square root law in log–log diagrams. The numbers corresponding to the parallel lines of slope 0.5 represent the values of the constant K.

Bakker [18] and Parrott [19]. The three models have the concrete requirements and climatic (humidity) loads as input factors.

Tuutti [14] based his method on the known diffusion theory of 'moving boundaries', which offers the following expression for the calculation of the rate of advance of the carbonation front:

$$\frac{C_s}{C_x} = \sqrt{\pi} \left[\frac{x/\sqrt{t}}{2\sqrt{D}} \right] \exp \left[\frac{x^2}{4Dt} \right] \operatorname{erf} \left(\frac{x/\sqrt{t}}{2\sqrt{D}} \right), \tag{10}$$

where C_s is the CO_2 concentration in the atmosphere (mol/kg), C_x is the amount of bound CO_2 (cement phases plus pore solution) (mol/kg), D is the CO_2 diffusion coefficient (m²/s), x is the carbonation depth (mm) and t is the time (s).

Bakker [18] based his proposal on a diffusion solution of the first order, but taking into account the internal moisture content of the concrete due to the RH cycling. This leads to the introduction of the concept of 'effective time' of action of the carbonation, as it is known that carbonation cannot progress in wet concrete. The expression reached is:

$$x_n = \sqrt{\frac{2D_c}{a}(C_1 - C_2) \left[t_{d1} + t_{d2} \left(\frac{x_1}{B} \right)^2 + t_{d3} \left(\frac{x_2}{B} \right)^2 + L + t_{dn} \left(\frac{x_{n-1}}{B} \right)^2 \right]}, \tag{11}$$

where

$$B = \sqrt{\frac{2D_v}{b}(C_3 - C_4)}, \tag{12}$$

$t_{d1}, \ldots, t_{dn}$ are the times of test, x_i is the carbonation depth (mm), a is the CaO content in the cementitious materials (mol/kg), b is the amount of water which evaporates from concrete (kg/m^3), D_c is the diffusion coefficient of CO_2 at a particular RH in concrete (m^2/s), D_v is the diffusion coefficient of water vapour at a particular RH in concrete (m^2/s), $C_1 - C_2$ is the difference of CO_2 concentration between air and concrete (mol/kg) and $C_3 - C_4$ is the difference in RH between air and concrete (mol/kg).

Parrott [19] followed a different approach offering an empirical solution based on the gas permeability of the concrete. Thus, he obtained an 'empiric expression' by mathematical fitting of carbonation depths measured in real structures. His fitting considers the 95% confidence range and the expression reached in function of the gas permeability is:

$$x = 64\frac{k^{0.4}t^{0.5}}{c^{0.5}}, \tag{13}$$

where x is the carbonation depth (mm), k is the oxygen permeability coefficient (m^2/s), t is the time (years) and c is the CaO content in the cement (mol/kg).

When applying eqns (10–13) to real data obtained from carbonated concrete, they give very similar results, as shown in Fig. 17. The choice of preference depends on the available input data.

2.3.4.2 Chloride penetration rates.   Chloride ingress through concrete cover can occur in marine or salty environments. When the chlorides arrive to the reinforcement in a certain quantity, local rupture of the passive layer is produced and corrosion starts. The process of penetration can be modelled as a pure diffusion process or by the square root law where V is the chloride penetration rate V_{Cl} [20]. In this case, the threshold of chlorides has to be pre-established as a front (Fig. 18) whose advance indicates the penetration depth of chlorides inducing corrosion. Plots similar to Fig. 17 can be then used.

For modelling chloride ingress other expressions based on Fick's second law are however more used. The solution more known is based on assumption of a semi-infinite media with a constant surface concentration, C_s [21].

$$C_x = C_s\left[1 - \text{erf}\frac{x}{2\sqrt{Dt}}\right], \tag{14}$$

where C_x is the chloride concentration at the depth x and time t and D is a non-steady state diffusion coefficient.

The current practice applies this equation to specimens tested in the laboratory in contact with a pond filled with NaCl solution (Fig. 19) or in cores drilled from real structures located in marine environments or in contact with deicing salts.

In the specimens or cores, slides are cut or ground parallel to the surface in contact with the chlorides. The equation is fitted, as Fig. 20 shows, to the graph of chloride concentration–penetration depth. By this fitting, the values of D and C_s are obtained. The values of D are then used to predict further evolution of the chloride profile. At this respect it has to be taken into account that, as chlorides react with the aluminates of the cement paste, the process is slower than a pure diffusion or absorption. And the Diffusion coefficient has to consider the amount

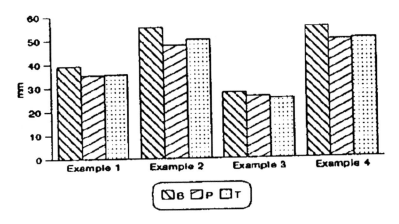

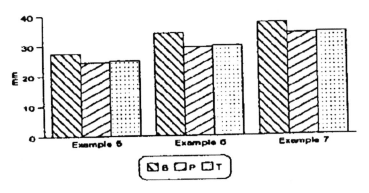

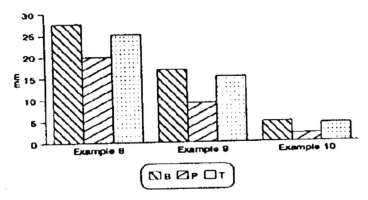

Figure 17: Carbonation depth results in concrete specimens obtained by means of the formulae proposed by Bakker (B), Parrot (P) and Tuutti (T) for the carbonation rate.

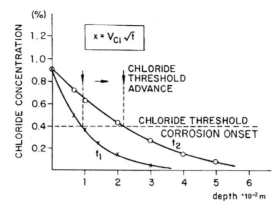

Figure 18: Definition of a chloride threshold enables to apply the square root law to chloride profiles. V_{Cl} = chloride penetration rate.

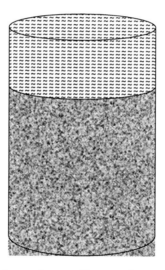

Figure 19: Typical concrete specimen with a pond containing a chloride solution in order to test the chloride diffusion coefficient.

of reacted or bound chlorides. In general, diffusion and reaction are not separated and an "Apparent D, D_{ap}" is used. Then in the following $D = D_{ap}$.

In spite of the general use of this model, it fails in many observations in real structures as its initial and boundary conditions are not fulfilled in many real environments. The main difficulties for its application are [22]:

- C_s does not remain constant and as the D value depends on it, both parameters should be given together. Thus, Fig. 21 shows two profiles (B and C) with different C_s but the same C_x at x in the rebar surface. B exhibits an apparent D

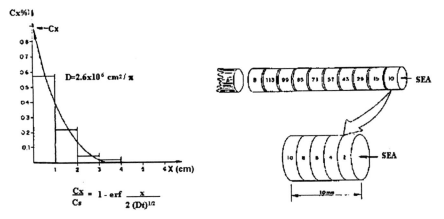

Figure 20: Fitting of the solution of Fick's second law to a chloride profile in order to obtain the diffusion coefficient and the surface concentration.

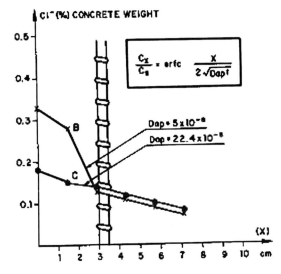

Figure 21: Two chloride profiles which indicate that the diffusion coefficient may be a misleading parameter. D_{ap} = apparent coefficient diffusion.

lower than that of C while both have the same risk for the reinforcement or even B supposes a higher one due to the higher total amount of chlorides.

- The D value is variable as well. It seems to decrease: with time, with higher external concentrations and with lowers temperatures. A time dependence of t has been introduced through [23]:

$$D(t) = D(t_0)\left[\frac{t}{t_0}\right]^n \qquad (15)$$

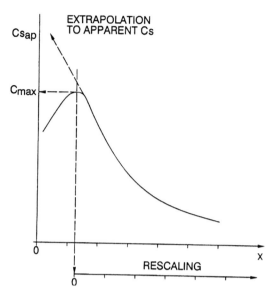

Figure 22: Pattern of chloride profiles showing a maximum beyond concrete surface. $C_{s\,ap}$ = apparent surface concentration.

where t_0 is the initial time, t is any time and n is an exponent of value around 0.5. Although the predictions made with this time dependence are not so unrealistic, the power n is so predominant that neglects any other influence.

- The error function equation does not take into account any mechanism other than diffusion, when absorption is also of very great importance in many real environments. In fact, the skin of the concrete surface behaves usually different than concrete bulk due to: (a) the direct contact with environment developing gradients of moisture from outside to the concrete interior; (b) its usual different composition with higher amount of paste and mortar and therefore higher porosity compared with the interior; and (c) the progressive carbonation in the cases where it occurs. The skin of the concrete then may exhibit a different D value than the interior modifying the profile and showing a maximum in the chloride concentration as the figure indicates. Also wet and dry regimes can lead into similar shapes of chloride profiles. The solution of this type of profile has been modelled through a 'skin effect' [24] although the D value can be also obtaining by rescaling the depth axis to placing the 'zero' at the maximum which would be C_s (Fig. 22).

2.3.5 The maximum tolerable amount of corrosion

Regarding the *maximum tolerable amount* of corrosion (loss of cross section of bars or "attack penetration", P_x) [14] in the case of new structures, both carbonation and chloride attack have to be considered separately. While in the case of carbonation a certain propagation period can be considered as part of the design service life (as the expected corrosion is homogeneous), almost no propagation period should be considered in the case of localised attack due to chlorides. This is because the uncertainties linked to how much localised and deep the corrosion pits may be do not enable at present the acceptance of such a risk from the design phase of the structure.

With regard to a limit for the case of the carbonation attack, a general suggestion is to accept a corrosion penetration depth of about 100–200 μm. Assuming a corrosion rate of about 5 μm/year, this would aim at propagation periods of 20–40 years [25]:

$$t_1 = \text{service life} = t_i + t_p = K_c\sqrt{t} + \frac{PL}{CR}, \tag{16}$$

where t_i is the corrosion initiation time (years), t_p is the corrosion propagation time (years), K_c is the carbonation coefficient, PL is the penetration limit (mm) and CR is the corrosion rate (I_{corr}, μm/year).

In the case of chlorides, the local attack may result in a very significant cross-section loss without negligible loss of metal. If the pits are very localized, the maximum propagation period to be considered should be 5–10 years [25].

2.3.6 Propagation period

When the reinforcement starts to corrode, several damages occur. The four main consequences of corrosion are illustrated in Fig. 23: (a) a loss of steel cross section due to the penetration of corrosion attack; (b) a loss in steel ductility due to the embrittlement caused by the corrosion inducing local acidification at the steel–concrete interface; (c) loss of steel–concrete bond due to the reduction of bar cross-section and the generation of iron oxides; and (d) cracking of concrete cover due to the pressure resulting from the production of oxides of higher volume than the parent steel [25, 26].

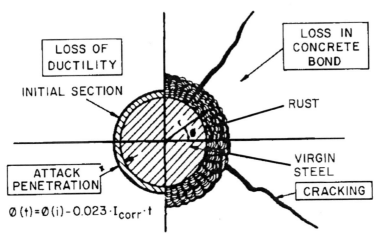

Figure 23: Consequences of reinforcement corrosion leading to loss of load-bearing capacity.

These damages affect to the load-bearing capacity of the whole structure and, in consequence, compromise the aesthetics, serviceability and safety, that is, the durability inherent to reinforced concrete as a construction material.

2.3.6.1 Detailed assessment of concrete structures affected by rebar corrosion. The main objective of a structural assessment is the residual safety level determination, in order to establish an adequate intervention programme with the higher degree of available information. On the other hand, a structural assessment can also be used for calibration of more simplified methodologies [26–28].

The three main aspects to be analysed in a structural assessment are:

- action, or better action effect, evaluated on the structure;
- deterioration process evaluation;
- safety or serviceability limit states verification.

The last two aspects will be explained below for structures affected by reinforcement corrosion:

Deterioration process evaluation. The attack by corrosion will be appraised by using a *penetration attack*, P_x, which is the loss of reinforcement radius. P_x is the main parameter that will allow a correlation with the general effects previously mentioned on the composite concrete–steel section [25]. This parameter can be measured visually from the residual diameter or estimated by means the corrosion rate I_{corr}.

P_x is related to I_{corr} by the expression:

$$P_x[\text{mm}] = 0.0116\alpha I_{corr}t,\qquad(17)$$

where t is the time and α is the pitting factor which takes into account the type of corrosion (homogeneous or localised) [29, 30].

The determination of I_{corr} in real structures will depend on several factors, and several strategies may be used for its determination in order to obtain a *representative value* of the loss of diameter P_x [28]:

- Several measurements at different times in different environmental conditions, to obtain a mean value to be used in the calculation; this can be achieved by using embedded sensors.
- A single value which is averaged with I_{corr} values obtained in a drilled core submitted in the lab to water saturation.
- The use a classification of exposure classes where representative corrosion values are proposed.

Reinforcement corrosion provokes on concrete–steel sections the effects shown in Fig. 23. These effects are related to the P_x value according the following principles:

Effective steel section reduction: Depending on the type of aggressive medium and the type of corrosion, their influence on the effective steel cross section is quite

Table 2: β values for crack width evolution.
Characteristic values.

	Top rebars	Bottom rebars
β	0.01	0.0125

different [25]. Thus, $\alpha = 2$ in eqn (17) if the corrosion is homogeneous and $\alpha = 10$ for very localized chloride-induced corrosion [28]. The calculation of the rebar diameter can be achieved through:

$$\Phi_{Res} = \Phi_0 - P_x = \Phi_0 - 0.0116 \alpha I_{corr} t, \tag{18}$$

where Φ_{Res} is the residual diameter and Φ_0 is the initial diameter.

Cover cracking: The oxides generated in the corrosion process provoke a tensional state in the concrete cover that will produce cover cracks of different widths, w, reducing consequently the cross-section of the concrete element and therefore their load-bearing capacity [31]. Several empirical expressions have been developed that can evaluate the crack width of the cover as a direct function of the corrosion attack P_x, and several geometric and mechanical parameters. The expression that is suggested is:

$$w[\text{mm}] = 0.05 + \beta[P_x - P_{x0}] \quad [w < 1\,\text{mm}]. \tag{19}$$

The value of P_{x0} depends mainly on the concrete cover/rebar diameter ratio, C/Φ, and the tensile strength of the concrete $f_{Ct,sp}$. Equation (24) provides an estimation of P_{x0}:

$$P_{x0} = 83.8 \times 10^{-3} + 7.4 \times 10^{-3}\left(\frac{C}{\Phi}\right) - 22.6 \times 10^{-3} f_{Ct,sp}, \tag{20}$$

where β depends on the rebar position in the element according to Table 2.

Loss of bond: The concrete–steel bond is the responsible of the composite behaviour of both materials [28]. However, corrosion provokes a reduction in bond due to the cover cracking and stirrups corrosion. Finally a limit state of bond can be achieved. Three main aspects are considered:

Residual bond assessment: Table 3 shows empirical expressions obtained based on several tests that allow the determination of realistic residual bond values. All of them depend on the attack penetration P_x.

For intermediate cases where the amount of stirrups is low, below the actual minimum, or the capacity of stirrups is strongly reduced by the corrosion effect,

Table 3: Relationship between P_x and residual bond strength f_b (MPa). Characteristic values.

With stirrups	No stirrups
$f_b = 4.75 - 4.64\,P_x$	$f_b = 2.50 - 6.62\,P_x$

Table 4: Bond strength values, intermediate amount of stirrups. Characteristic values.

f_b	$10.04 + m(1.14 + P_x)$
m	$-6.62 + 1.98(\rho/0.25)$
ρ	$n[(\phi_w - \alpha P x)/\phi]^2$

Where ϕ is the initial longitudinal diameter (mm), ϕ_w is the transversal diameter (mm), n is the number of transversal reinforcements, α is the coefficient that depends on the type of corrosion, m is the adjust parameter and ρ reinforcement ratio.

the expressions in Table 4 may be applied. These expressions are applied for P_x values between 0.05 and 1 mm with $\rho \leq 0.25$.

Influence of external pressures that can be present due to external supports: Thus, expressions similar to that presented in Eurocode 2 have been developed:

$$f_b = \frac{(4.75 - 4.64 P_x)}{1 - 0.098p}, \tag{21}$$

where f_b is the bond strength (MPa), P_x is the corrosion attack (mm) and p is the external pressure in the bond zone (MPa). This expression can be used for the bond strength evaluation of the rebar at element ends.

Figure 24 shows an application of these expressions in a reinforcing bar diameter of 20 mm without stirrups (curve 3) or with $4\phi8$ stirrups. Curves 1 and 2 correspond to the bond reduction with a reduction at the end of the element without pressure (1 with homogeneous corrosion and 2 with pitting corrosion), curve 4 corresponds to an external force of 5 MPa.

Relationship between bond and crack width: Several expressions have been developed for relating the residual bond with the crack width (Table 5).

Rebar ductility: Several tests of corroded rebars have shown an important reduction in the rebar ductility not only in the final strain of the rebar but also in the strain–stress curve of the corroded rebar. The tests show that the yield point is

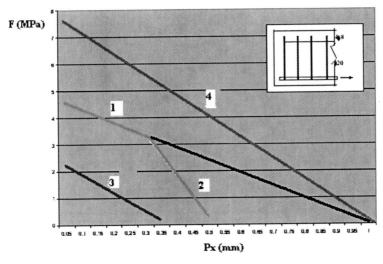

Figure 24: Loss of bond as a function of P_x (in mm). (1) Element with stirrups without pits; (2) element with pits in stirrups; (3) element without stirrups; (4) element with stirrups and external pressure.

Table 5: Relationship between residual bond strength f_b (MPa) and crack width w (mm). Characteristic values.

Stirrups	No stirrups
$f_b = 4.66 - 0.95w$	$f_b = 2.47 - 1.58w$

diffuse and the final strain is considerably reduced. However, in all cases the final value of the ultimate strain is above 1% which is the value to be used in Ultimate Limit State.

Structural analysis. The structural analysis can be carried out following the main principles of structures with linear–elastic analysis. It is important to identify the presence of cracks in sections produced by corrosion or not in order to reduce their stiffness in the structural model.

Limit state verification. *Ultimate limit state:* For slab and beams, a conservative value of the ultimate bending moment can be calculated by using the classical models adding correction in order to take into account the reduced steel section and the concrete section spalled [28]. A possible reduction due to bond deterioration as a result of corrosion should be considered, especially if the corrosion attack is on the tensile zone of the beams.

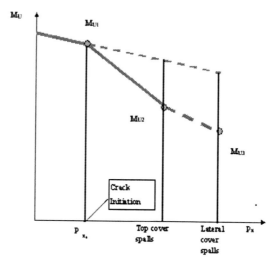

Figure 25: Time–history evolution of ultimate bending moment.

Although shear and bending moment are assumed to have the same safety in the design phase for beams without corrosion because the shear formulation is considered to be conservative, several factors may induce a premature failure of shear for corroded beams, such as:

- small diameters on stirrups,
- lower cover for stirrups,
- spalling of cover.

In order to check the ultimate axial effort of a column element, the reduction should be applied on the concrete section in the case of spalling, and if there are no stirrups, on the longitudinal bars subjected to compression due to risk of buckling.

Although no tests have been performed regarding punching shear it is possible to extrapolate the shear test to slabs in order to use a verification procedure for punching in slabs.

The time–history evolution of the load-bearing capacity is simplified by using a linear interpolation between the crack initiation and the spalling point of lateral and top covers. It can be proved that, for normal concrete sections the exact calculation (step by step) provide a negligible error using this simplification. Thus, it is necessary to calculate three points in Fig. 25. The ultimate section effort (M_{U1} in the example) at the crack initiation point P_{x0}, the ultimate section effort (M_{U2}) when the top cover spalled (it is assumed that the top cover can be neglected when the crack width at top is >0.2 mm), and finally (M_{U3}) when the lateral cover spalls (the same criteria applies).

Serviceability limit state: The serviceability limit states to be checked should be [28]:

- exterior aspect of the structures (rust, spalling);
- cracking of the cover due to corrosion or excessive loading;
- excessive deflections.

For the deflection and crack checking due to loading, the expressions provided by Eurocode 2 can be used, reducing the steel section, the spalled concrete and the cracking if it exists. However, it is the owner of the structure who has to establish their acceptable degree for their structures regarding serviceability conditions.

3 Final remarks

Reinforced concrete is the most widely used construction material. Cement consumption ranges from 0.2 to 1 tonne per person, which means 1.4–7 tonnes of concrete per person. With more than a century of experience, it is now known that it performs adequately in very diverse climates and moderately aggressive environments [32].

However, for very long life, beyond 75 years, or standing up to very aggressive environments, reinforced concrete needs a minimum quality and cover thickness, or even supplementary protection methods to avoid reinforcement corrosion.

In sum, concrete structures should be designed with a specified maintenance regime and periodically inspected in order to lengthen their service life. Management tools are being developed to help the owners to protect their structures and contribute to reduce repair costs.

Nomenclature

E actual potential
E_0 standard potential
R gas constant
T absolute temperature
e^- electrons exchanged
k equilibrium constant
I current applied or recorded
i_0 exchanged current
η change in potential, overpotential ($E_c - E_a$)
R electrical resistance
E_a anodic potential
E_c cathodic potential
R_p polarization resistance
B   constant based on Tafel slopes
b_a anodic Tafel slope
b_c cathodic Tafel slope
t_l service life
t_i initiation period
t_p propagation period

x	attack penetration depth
t	time
K	constant depending on concrete and ambient characteristics
C_s	CO_2 concentration in the atmosphere
C_x	amount of bound CO_2
D	CO_2 diffusion coefficient
D_c	diffusion coefficient of CO_2 at a particular RH in concrete
D_v	diffusion coefficient of water vapour at a particular RH in concrete
t_d	time of test
a	CaO content in cementitious materials (Bakker model)
b	amount of water which evaporates from concrete
$C_1 - C_2$	difference of CO_2 concentration between air and concrete
$C_3 - C_4$	difference in RH between air and concrete
c	CaO content in cementitious materials (Parrott model)
V_{Cl}	chloride penetration rate
D	non-steady state diffusion coefficient
t_0	initial time
K	carbonation coefficient
PL	penetration limit
or I_{corr}	corrosion rate (CR)
P_x	penetration of attack
α	pitting factor
Φ_{Res}	residual diameter
Φ_0	initial diameter
$f_{Ct,sp}$	tensile strength of the concrete
w	crack width
β	parameter for the estimation of crack width that depends of the rebar position
f_b	residual bond strength
ϕ	initial longitudinal diameter in mm
ϕ_w	transversal diameter in mm
n	number of transversal reinforcements
α	coefficient that depends on the type of corrosion
m	adjust parameter
ρ	reinforcement ratio
p	the external pressure in the bond zone

References

[1] Evans, U.R., *The Corrosion and Oxidation of Metals*, Arnold and Co.: London, 1960.

[2] Tomashov, N.D., *Theory of Corrosion and Protection of Metals*, Macmillan Co.: New York, 1966.

[3] Pourbaix, M., *Atlas of Electrochemical Equilibria in Aqueous Solutions*, NACE: Houston, TX, 1974.

[4]   Uhlig, H.H., *Corrosion and Corrosion Control*, Wiley and Sons: New York, 1971.

[5]   Stern, M. & Geary, A.L., Electrochemical polarization. A theoretical analysis of the shape of polarization curves. *Journal of the Electrochemical Society*, **104(1)**, pp. 56–63, 1957.

[6]   Andrade, C. & Gonzalez, J.A., Quantitative measurements of corrosion rate of reinforcing steels embedded in concrete using polarization resistance measurements. *Werkst. Korros.*, **29**, p. 515, 1978.

[7]   Feliú, S., González, J.A., Feliú, S., Jr. & Andrade, C., Confinement of the electrical signal for in situ measurement of polarisation resistance in reinforced concrete. *ACI Materials Journal*, September–October, pp. 457–460, 1990.

[8]   Slater, J.E., *Corrosion of Metals in Association with Concrete*, ASTM STP-818, ASTM: Philadelphia, PA, 1983.

[9]   Hoar, T.P., Passivity, passivation, breakdown and pitting. *Corrosion*, Vol. 1, ed. L.L. Shreir, Butterworths: London, 1978.

[10] Hausman, D.A., Criteria of cathodic protection of steel in concrete. *24th Conf. of NACE*, Cleveland, OH, 1968.

[11] Alonso, C., Castellote, M. & Andrade, C., Chloride threshold dependence of pitting potential of reinforcements. *Electroquimica Acta*, **47**, pp. 3469–3481, 2002.

[12] Pedeferri, P., *Corrosione e protezione dei materially metallici*, Clup.: Milano, 1981.

[13] Galvele, J., A stress corrosion cracking mechanism based on surface mobility. *Corrosion Science*, **27**, pp. 1–3, 1987.

[14] Tuutti, K., *Corrosion of Steel in Concrete*, CBI Research Report 4:82, Swedish Cement and Concrete Research Institute: Stockholm, 1982.

[15] Gjorv, O.E. & Vennesland, O., Diffusion of chloride ions from seawater into concrete. *Cement and Concrete Research*, **9**, pp. 229–238, 1979.

[16] Page, C.L., Short, N.R. & El Tarras, A., Diffusion of chloride ions in hardened cement pastes. *Cement and Concrete Research*, **11**, pp. 395–406, 1981.

[17] Andrade, C. & Alonso, C., Progress on design and residual life calculation with regard to rebar corrosion of reinforced concrete. *Techniques to Assess the Corrosion Activity of Steel Reinforced Concrete Structures*, ASTM STP-1276, eds. N. Berke, E. Escalante, C.K. Nmai & D. Whiting, pp. 23–40, 1996.

[18] Bakker, R., Prediction of service life reinforcement in concrete under different climatic conditions at a given cover. *Int. Conference on Corrosion and Corrosion Protection of Steel in Concrete*, Sheffield, UK, ed. R.N. Swamy, 1994.

[19] Parrott, L.J., Design for avoiding damage due to carbonation-induced corrosion – ACI-SP-145-15. *Int. ACI/CANMET Conference of Durability of Concrete*, Nice, France, ed. M. Malhotra, pp. 283–298, 1994.

[20]   Andrade, C., Sagrera, J.L. & Sanjuán, M.A., Several years study on chloride penetration into concrete exposed to Atlantic Ocean water. *2nd Rilem Workshop of Testing and Modelling Chloride Penetration into Concrete*, Paris, 2000.

[21]   Collepardi, M., Marcialis, A. & Turriziani, R., Penetration of chloride ions into cement pastes and concretes. *Journal of the American Ceramic Society*, **55**, pp. 534–535, 1972.

[22]   Andrade, C., Díez, J.M., Alonso, C. & Sagrera, J.L., Prediction of time to rebar initiation by chlorides: non-Fickian behaviour of concrete cover. *Concrete in the Service of Mankind*, eds. R.K. Dhir & P.C. Hewlett, E & FN Spon: London, pp. 455–464, 1996.

[23]   Mangat, P.S. & Molloy, B.T., Prediction of long term chloride concentration in concrete. *Materials and Structures*, **27**, pp. 338–346, 1994.

[24]   Andrade, C., Díez, J.M. & Alonso, C., Mathematical modelling of a concrete surface 'skin effect' on diffusion in chloride contaminated media. *Advanced Cement Based Materials*, **6**, pp. 39–44, 1997.

[25]   Andrade, C., Alonso, C. & Rodriguez, J., Remaining service life of corroding structures. *IABSE Symposium on Durability*, Lisbon, September, pp. 359–363, 1989.

[26]   Rodriguez, J., Ortega, L.M. & García, A.M., Assessment of structural elements with corroded reinforcements. *Int. Conference on Corrosion and Corrosion Protection of Steel in Concrete*, Sheffield, UK, ed. R.N. Swamy, pp. 171–185, 1994.

[27]   Rodriguez, J., Ortega, L.M., Casal, J. & Díez, J.M., Assessing the structural condition of concrete structures with corroded reinforcements. *International Congress on Concrete in the Service of Mankind*, Dundee, UK, Vol. 5, June, eds. R. Dhir and P.C. Hewlett, 1996.

[28]   Rodriguez, J., Andrade, C., Izquierdo, D. & Aragoncillo, J., Manual of assessment of structures affected by reinforcement corrosion, EU Project Contecvet IN309021, http://www.ietcc.csic.es/

[29]   González, J.A., Andrade, C., Alonso, C. & Feliú, S., Comparison of rates of general corrosion and maximum pitting penetration on concrete embedded steel reinforcement. *Cement and Concrete Research*, **25(2)**, pp. 257–264, 1995.

[30]   Andrade, C. & Arteaga, A., *Statistical Quantification of the Propagation Period*, Internal report Brite/Euram BE95-1347, Duracrete, 1998.

[31]   Alonso, C., Andrade, C., Rodríguez, J. & Díez, J.M., Factors controlling cracking of concrete affected by reinforcement corrosion. *Materials and Structures*, **31**, pp. 435–441, 1998.

[32]   Andrade, C., Alonso, C. & Sarría, J., Corrosion rate evolution in concrete structures exposed to the atmosphere. *Cement and Concrete Composites*, **24**, pp. 55–64, 2002.

CHAPTER 7

Environmental pollution effects on other building materials

P. Rovnaníková

*Department of Chemistry, Faculty of Civil Engineering,
Brno University of Technology, Brno, Czech Republic.*

1 Introduction

In building materials, water is present in two forms: free water and bound water. Free water is a part of moisture in materials whose physical properties are comparable with pure water. Bound water is a part of water whose properties differ significantly from free water. The water of crystallisation and the water that reacts with building materials are also classified as bound water. These types of bonded water can be the reason for the damage of building materials.

The bonding of water in a material can be chemical, physical-chemical, or physical. The chemical bond is the most stable one. Physical-chemical bonds are characteristic of water bonding to the material surface. The strength of the bond is affected by the structure of the surface in a significant way. The most typical representative of physical-chemical bonding is the hydrogen bond. Physically bound water exhibits the weakest bond and the lowest interaction with the molecules of a solid material.

The moisture contained in building materials derives from three sources: rainwater, water rising through capillary action, and condensed water. The quantity of water that attacks building materials is partly influenced by the choice of the construction site.

The quantity of *rainwater* penetrating into building materials depends not only on the total rain quantity but also on its intensity, air flow, and the properties of materials and their surfaces. The wind promotes rainwater penetration into building materials but also increases water evaporation from the surface, i.e. desiccation from the walls and external coatings. The porosity of the materials used and the form of the type of construction influence the humidification of materials. Certain places on façades, such as ledges with a small slant, serve as a water reservoir. Rainwater usually contains dissolved corrosive pollutants that are transported through the coating into the construction materials and cause their deterioration.

Water rising via *capillary action* always contains indispensable amounts of dissolved salts such as sulphates, nitrates, chlorides, and ammonium cations. The quantity of water penetrating into walling depends primarily on the level of groundwater but also on the character of the subsoil material. The elevation of water is determined by a building material's properties and the characteristics of its surface because they may influence water evaporation.

Theoretically, the height to which a liquid rises in a narrow capillary is given by the mathematical formula:

$$h = \frac{2\sigma \cos \Theta}{r \rho g}, \qquad (1)$$

where r is the radius of the capillary, ρ is the water density, g is the acceleration due to gravity, σ is the surface tension, and Θ is the angle of wetting. For example, when the capillary radius size is 0.1×10^{-3} m, the balanced water level is 0.149 m, but in the case of a capillary with radius 0.01×10^{-3} m, the theoretical height of the water level is 1.490 m. The result of the calculation of water level height is only theoretical because the internal geometry of a porous body is very complicated. Inorganic building materials contain pores with radii from 10^{-9} to 10^{-3} m. In addition to the capillary size, the quantity of water that infiltrates into the underground of building surroundings also has an appreciable influence on humidification of building materials.

Condensed water originates in the atmosphere. The absolute humidity Φ of air is defined by the equation:

$$\Phi = \frac{m}{V}, \qquad (2)$$

where m is the mass of water vapour in kg and V is the volume of wet air in m^3.

The relative humidity of air φ' is the ratio of the absolute humidity Φ of the air to the humidity of the air saturated by water vapour at the specified temperature and pressure Φ_{nas}:

$$\varphi = \frac{\Phi}{\Phi_{nas}} \times 100 \qquad (3)$$

The relative humidity φ of completely dry air is 0%; the humidity of water vapour saturated air is 100%.

The limiting value of relative humidity φ that can be achieved in a capillary of radius r is determined by the Kelvin equation [1]:

$$\varphi = \exp\left(-\frac{2\sigma M}{\rho_l R_g T} \frac{1}{r}\right) \qquad (4)$$

where M is the molar mass of water, R_g is the universal gas constant, σ is the surface tension on the gas–solid interface, ρ_l is the pressure in the liquid phase, T is the temperature, and r is the capillary radius. If the amount of water vapour in the capillary exceeds φ, condensation begins; this phenomenon is denoted as *capillary condensation*. The Kelvin equation expresses the fact that the partial pressure of saturated water vapour decreases with the increasing surface curvature, i.e. with the decreasing radius of a capillary.

In what kinds of pores is capillary condensation of practical importance? When the numerical values of the Kelvin equation are installed, the range of capillary radius where capillary condensation arises is obtained. Condensation begins somewhere at 1 or 2 nm and ends at 50 nm of the capillary radius. For a smaller capillary radius, surface adsorption is the dominant mechanism of water sorption. For capillaries of larger radius, only the water vapour transport mechanism may occur. When the building material is in direct contact with liquid water, capillary suction may occur.

2 Lime and mixed external coating

External coating may be regarded as a part of the construction that has no load-bearing function but that is very important from the viewpoint of the protection of structural materials. Moreover, coatings influence the architectural appearance of buildings.

Plasters and renders have two functions:

- protection and thermal insulation, and
- aesthetics.

The application of coatings was probably conditioned by the requirement of protecting the walling materials against the action of atmospheric factors such as rain and dissolved chemicals, like carbon dioxide, sulphur dioxide, etc. If damaged, plasters may easily be renewed while the damaged walling material is replaced only with difficulty. Based on the kind of walling material used, damage might result in its total destruction.

From the point of view of thermal insulation, coatings containing a higher number of pores find a greater application. These coatings include mortars based on non-hydraulic lime because the porosity of these mortars, compared with the lime-cement mortars or cement mortars, is considerably higher. Both from history and present experience, we know special kinds of thermal insulation coatings in which pores are purposefully created.

2.1 Composition and properties

The external coating of buildings is composed of binders; sand and sometimes admixtures are also used. The main binders for mortars used for rendering are lime, cement, and pozzolanic materials. Mixed binders like lime cement or lime pozzolana are also very often used.

Lime mortars for coating are made of lime in the form of lime putty or powdered hydrated lime, sand, and water. The solidification process is called carbonation because calcium hydroxide reacts with carbon dioxide and produces calcium carbonate. The carbonation process is slow, and it depends on the concentration of carbon dioxide in the air, the relative humidity of the air, and the porosity of the mortar structure. The process of stable structure formation consists of two partial steps. First, porous building materials such as bricks suck out the mixing water from mortar. Second, carbonation takes place. The reaction proceeds according to the following reaction:

$$CO_2 + Ca(OH)_2 + H_2O \rightarrow CaCO_3 + 2H_2O$$

This is a long-term process that is controlled by the diffusion of carbon dioxide through the layer of coating. The pH value subsequently decreases from the surface to the brick or stone masonry. The fully carbonated lime mortar coating reaches a pH value of 8.3.

Coatings based on hydraulic lime form a stable structure dependent on the hydraulic modulus of lime, M_H. The hydraulic modulus is characterised by the ratio $M_H = CaO/SiO_2 + Al_2O_3 + Fe_2O_3$. Strongly hydraulic limes with hydraulic modulus $M_H = 1.7–3.0$ have a large portion of hydraulic compounds and a small quantity of free lime, whereas poorly hydraulic limes with $M_H = 6.0–9.0$ contain a large quantity of free lime. The formation of a stable structure happens in two steps: at first, the hydraulic compounds react with water to form hydrated calcium silicates and aluminates. Subsequently, the carbonation of free lime commences. The products of hydration are less soluble in water and more resistant to the action of the corrosive environment than the original reactants.

The solidification and hardening of *cement mortar coatings* are the physical demonstrations of the hydration of clinker minerals. Anhydrous compounds react with water to form hydration products. The chemical and physical bonds in calcium silicates and aluminates manifest themselves as compressive and bending strengths.

Mortars for external coatings are usually formed of a mixed (or composite) binder like lime-cement or lime-pozzolana. The properties of *lime-cement binder coatings* are similar to those of hydraulic lime mortars. *Lime-pozzolana coatings* are known to have been used for centuries. Pozzolanic materials contain amorphous silicon dioxide and/or amorphous aluminosilicates, which react with calcium hydroxide in fresh mortar to form compounds similar to those in hydraulic lime mortar. Therefore, lime-pozzolana external coatings are more resistant than coatings based on lime mortar [2].

2.2 Binding materials damage

The deterioration of coatings according to the nature of the damaging action is classified into three groups:

- physical deterioration – mechanical influences (impacts), high and low temperatures, crystallisation pressures;
- chemical deterioration – gas pollutants, aerosols of acids, hydroxide and salts solutions, organic substances; and

Table 1: Main corrosive gases, aerosols and ions in environment of buildings.

Component	Chemical base
Gases	CO_2, SO_2, SO_3, NO_x, H_2S, HCl
Aerosols	Salts, acids
Ions in rain and ground water	SO_4^{2-}, Cl^-, NO_3^-, CO_{2agr}, Mg^{2+}

- biological deterioration – compression of plant roots, chemical reaction of animal excrements, action of micro-organisms.

External coatings are affected by the action of weather and aggressive substances from the environment, from the air, and from capillary action water in which the aggressive compounds are dissolved. The dissolved aggressive substances can be transported through the masonry and the evaporating zone, or are directly transported to the coating. Binders used for rendering and pointing mortars are easily deteriorated. Typically, the soluble products formed by the reactions between the binder and compounds in water can be leached out from the coating by rainwater. The most common corrosive substances that are present in the atmosphere, in the ground water, and are dissolved in rainwater are given in Table 1.

Gases dissolved in rainwater and groundwater is corrosive to lime, limepozzolana, lime-cement, and hydraulic lime binders. The pollution of the atmosphere is not problematic only since the last several decades. As early as in 1661, London chronicler John Evelyn described sulphur dioxide pollution of the air in [3].

The acidic components of rainwater are responsible for the low pH value of rainwater. As of the year 2000, the most acidic rainfall recorded in the USA had a pH of about 4.3. Normal rain is slightly acidic because carbon dioxide dissolves in it, so it has a pH of about 5.5. Sulphur dioxide and nitrogen oxides (NO_x) are the primary causes of acid rain. In the USA about two-thirds of all SO_2 emissions and one-fourth of all NO_x emissions result from electric power generation that relies on burning fossil fuels. Very strict laws that limit the amount of aggressive substances are in force in Europe. Therefore, the rainwater in Europe has a higher pH value.

Groundwater transport to the masonry and to the coatings can proceed from the ground soil. Above all, sulphates, nitrates, chlorides, and ammonium salts are transported. Bricks often contain sulphates because these are present in the raw materials bricks are made of. The capillary water dissolves $CaSO_4 \cdot 2H_2O$ and $Na_2SO_4 \cdot 10H_2O$, and the ions are transported to the coating.

The presence of nitrates in plasters and renders is influenced by the action of bacteria that oxidise organic nitrogen to nitrates. Nitrates appear especially in the ground parts of coatings and in places where the coating is in contact with the horizontal surface where excrements of animals, mostly those of birds, are present. Nitrogenous substances leaching from the excrements infiltrate into coatings, and the nitrification bacteria oxidise them to nitrates. Nitrates are highly soluble in water and some of them easily crystallise as hydrates.

Chlorides may be found in coatings, especially in the vicinity of places where sodium chloride is used as a de-icing salt for roads and pavements. Then, salt is transported in the form of a solution to the ground-level parts of plasters and walling where it deposits; after water evaporation, the salt crystallises. Water–NaCl aerosol is present in the atmosphere of coastal areas and is also created by the passage of vehicles along wet roads treated with de-icing salt. These aerosols may also deposit in higher parts of facades. The concentration of NaCl deposits from aerosols is not too high, however, and is not usually the cause of the disintegration of coatings as is the case of ground-level parts.

Crystallisation and recrystallisation of salts represent some of the causes of plaster damage. The formed crystals exert crystallisation pressures on their surroundings and, if the pressure is higher than the plaster strength, failure occurs [4]. The crystallisation and recrystallisation pressures of one of these salts are illustrated in Tables 2 and 3.

If salts crystallise on the coating surface, mechanical damage need not occur. However, if crystallisation occurs in the coating pores, the coating disintegrates as a result of crystallisation pressure.

When the binder content of the coating is low and the limit of cohesion is exceeded, the coating decays. Salts also crystallise on the masonry surface under the coating, and this causes delamination of the coating from the masonry surface.

The durability of calcium carbonate (the main binding compound in lime and lime-mixed mortars) to acid corrosive environments is very low. The binder decomposes and usually forms soluble products that are leached from the coating by rainwater. Chemical reactions harmful to lime and lime-mixed coatings are described below.

Table 2: Crystallisation pressure of salts.

Compound	Crystallisation pressure (MPa)
$CaSO_4 \cdot 2H_2O$	28.2
$MgSO_4 \cdot 2H_2O$	10.5
$Na_2SO_4 \cdot 10H_2O$	7.2
$Na_2CO_3 \cdot 10H_2O$	7.8
NaCl	5.4

Table 3: Recrystallisation pressure of salts.

Initial salt	Final salt	Recrystallisation pressure (MPa)
$CaSO_4 \cdot 0.5H_2O$	$CaSO_4 \cdot 2H_2O$	160
$MgSO_4 \cdot 6H_2O$	$MgSO_4 \cdot 7H_2O$	10
$Na_2CO_3 \cdot H_2O$	$Na_2CO_3 \cdot 7H_2O$	64

Carbon dioxide dissolves in water but the carbonic acid forms only 1% of oxide. The remaining CO_2 is in the molecular form and, if water is soft (without dissolved ions), carbon dioxide is very aggressive. The reaction of aggressive CO_{2agr} produces calcium hydrogen carbonate $Ca(HCO_3)_2$:

$$CaCO_3 + CO_2 + H_2O \Leftrightarrow Ca^{2+} + 2HCO_3^-$$

The reaction is reversible and, under suitable conditions, $Ca(HCO_3)_2$ decomposes back to calcium carbonate. When enough liquid water is present, $Ca(HCO_3)_2$ is leached from the coating. The amount of binder in the coating decreases, and the coating decays.

Sulphur dioxide dissolves in water and partly forms sulphurous acid; sulphur trioxide forms acid as well. Both acids decompose lime and lime-mixed binders in coatings. The final result is formation of gypsum, $CaSO_4 \cdot 2H_2O$. The reaction can be expressed in a simplified manner as:

$$CaCO_3 + SO_2 + 1/2O_2 + 2H_2O \rightarrow CaSO_4 \cdot 2H_2O + CO_2$$

and

$$CaCO_3 + SO_3 + 2H_2O \rightarrow CaSO_4 \cdot 2H_2O + CO_2$$

Gypsum has a large molar volume and, under favourable humidity conditions, large crystals of gypsum form. As a result, crystallisation pressure decays the coating.

Hydrogen sulphide originates by the action of micro-organisms during the disintegration of organic substances and is released during volcanic activities. In aqueous solutions it behaves like a weak acid.

Nitric acid is formed from nitrogen dioxide, oxygen, and water. This acid decomposes calcium carbonate according to the reaction:

$$CaCO_3 + 2HNO_3 \rightarrow Ca(NO_3)_2 + CO_2 + H_2O$$

Calcium nitrate is a very soluble substance, and it is leached from the coatings by the action of rainwater.

Hydraulic binders are mostly damaged by the acid constituents in ambient air or by all types of aerosol solutions. In general, the degradation of the products of hydraulic and pozzolanic hardening proceeds similarly in all types of hydraulic binders. Calcium hydroxide that is formed during cement hydration will preferably undergo a reaction with the acid corrosive components. The chemical reactions are described as follows:

$$Ca(OH)_2 + 2HCl \rightarrow CaCl_2 + 2H_2O$$

$$Ca(OH)_2 + H_2SO_4 \rightarrow CaSO_4 \cdot 2H_2O$$

$$Ca(OH)_2 + 2HNO_3 \rightarrow Ca(NO_3)_2 + 2H_2O$$

In coatings where the binders are composed of products of the lime carbonation and of hydraulic reactions, the degradation commences as in lime coatings. Later on, the products of the hydraulic reactions in lime-cement plasters are attacked. Acids disintegrate both C-S-H and calcium aluminate hydrates.

Like lime-cement coatings, the coatings containing pozzolana are also more resistant to the action of corrosive substances from the ambient environment than coatings based on non-hydraulic lime. There are mortars preserved from the past that were produced more than two thousand years ago that reveal no signs of degradation. These 'long-term' mortars were made of lime and pozzolana admixtures. As stated by Davidovits [5], ancient mortars contained as much as 40 wt.% of substances of zeolitic character that formed through the reaction of lime and pozzolana admixtures. Lime zeolites of phillipsite type, $3CaO \cdot 3Al_2O_3 \cdot 10SiO_2 \cdot 12H_2O$, but also compounds containing alkali components, sodium or potassium, e.g. analcime, have been identified. These compounds are also found in nature and distinguish themselves by a high stability against the action of corrosive substances from the ambient environment.

Sabboni et al. [6] described the atmospheric deterioration of ancient and modern hydraulic mortars. They investigated the damage of mortars based on lime-cement, lime-pozzolana, and hydraulic lime collected in Belgium, Italy, and Spain from ancient and modern buildings. They found sulphation to be the main damage mechanism in all the analysed mortars. Gypsum was the primary and ettringite the secondary damaging product. Ettringite is derived from the reaction between the gypsum and the calcium aluminate hydrates present in the binder. Ettringite was formed only inside the samples, due to its instability with the atmospheric carbon dioxide.

The impact of sulphur dioxide on lime mortar and lime mortar with sepiolite was studied by Martínez-Ramírez et al. [7, 8]. For their experiments, they used an atmospheric simulation chamber with aggressive gases such as NO_x, SO_2, and oxidant O_3. They found the reaction rate between lime mortar and NO gas was low. NO_2 as a pollutant has a lower reactivity, but it slightly increases in the presence of water. Sulphur dioxide is the most reactive gas pollutant affecting lime mortar. The reactivity depends on the presence of water and ozone. When both water and ozone are present, the rate of the reaction significantly increases. Clay minerals with fibrous sepiolite reduce the reaction rate of SO_2 when added to lime mortar.

Frost damage of coating occurs in those parts of the facade that are frequently frozen while very wet. Resistance against damage initiated by the freezing of water is influenced by the pore structure of the coating mortar. There is a very close relationship between the pore structure of the hardened mortar and frost resistance [9].

2.3 Pollution prevention measures

Measures against harmful substances from the atmosphere depend on the situation whether the coating is on a new construction or whether this refers to really historical buildings. For new constructions lime-cement coatings are used, usually provided with a suitable paint on the surface. These paints have not only an aesthetic function but also a protective function. Façade paints used in the present times contain hydrophobisation substances that prevent the penetration of rainwater into the coating, and in this way, impede its damage. On the other hand, these paints must have a low resistance to diffusion (0.01–0.3 m), and not inhibit the diffusion of water vapour from the building material to the ambient environment.

Coatings without any paint or provided with a paint having no hydrophobic properties may be painted with a hydrophobisation inhibitor. These paints are based on silicon compounds that are bonded by the Si-O-Si chain to the sub-base, and the organic part of the molecule (CH_3- or C_2H_5-) that has hydrophobic properties, is orientated towards the atmosphere.

The comprehensive measures that must be taken against capillary water that often causes the damage of coatings in ground-floor parts of facades include:

- removing the source of moisture;
- additional horizontal insulation; and
- using mortar of higher porosity.

The first measure may be fulfilled if capillary water is the source of moisture, e.g. water from a damaged sewer, rainwater gutter, or the incline of the surrounding ground. In this case, such improvements may be carried out that minimise the source of capillary water.

The second measure may be achieved in two ways: either by inserting a band made of plastic or bituminous material into a gap made in the masonry with a saw, or with injection solutions pressed into holes drilled in advance that are a certain distance apart.

The above-mentioned measures prevent the capillary action of water in the masonry and the damage of the coating on its surface. Nevertheless, in the ground floor parts of the façade, a special kind of coating of a higher porosity is used (minimum 40% of pore volume in hardened mortar). This coating does not prevent the diffusion of water vapour from the masonry, and the existing pores serve for depositing the crystals of salts brought by water from the subsoil.

3 Plaster of Paris

3.1 Composition and properties

Gypsum plaster is a structural binder that has been in use since the ancient times. The Egyptians used gypsum plaster as a masonry mortar in the construction of pyramids, and the Phoenicians also used gypsum mortars. In the 17th century, gypsum plaster was used in Paris as a surface finish for timber structures that had to serve as a protection against fire. The King of France issued a writ concerning the protection of constructions after the great fire that struck and destroyed a large part of London in the year 1666. Till now, gypsum plaster used for renders and plaster coatings is called 'Plaster de Paris'.

Gypsum is a raw material for the production of gypsum plaster of the following chemical composition: calcium sulphate dihydrate, $CaSO_4 \cdot 2H_2O$. This mineral crystallises in the monoclinic system in the shape of small columns, tables or needles.

Gypsum is stable up to the temperature 42°C; at a higher temperature, the loss of chemically bonded water occurs [10].

Natural gypsum is greyish due to the presence of impurities. If pure limestone is used for the process of desulphurisation, the gypsum is white and contains as much as 98% $CaSO_4 \cdot 2H_2O$.

During the production of calcinated plaster, several products are made, depending on the temperature of firing. The production of calcinated plaster and gypsum binders is based on the ability of gypsum to release water when heated, according to the following reaction:

$$CaSO_4 \cdot 2H_2O \rightarrow CaSO_4 \cdot \tfrac{1}{2}H_2O + 1\tfrac{1}{2}H_2O \rightarrow CaSO_4 + 2H_2O$$

Calcium sulphate hemihydrate is the substance of quick-setting gypsum plaster that is in α-form or β-form. Anhydrites are produced during further heating, and at a temperature higher than 1000°C, $CaSO_4$ partially decomposes

$$CaSO_4 \rightarrow CaO + SO_3$$

and generates a solid solution ($CaO + CaSO_4$) [11]. The mixture of 75%–85% of anhydrite and 2%–4% of calcium oxide and up to 10% of other constituents is called slowly setting gypsum plaster.

The setting and hardening of gypsum plaster is a process that is the reverse of its production. When mixed with water, gypsum plaster is dissolved and gypsum subsequently crystallises according to the reaction:

$$CaSO_4 \cdot \tfrac{1}{2}H_2O + 1\tfrac{1}{2}H_2O \rightarrow CaSO_4 \cdot 2H_2O$$

With the hydration of anhydrite, the following reaction takes place:

$$CaSO_4 + 2H_2O \rightarrow CaSO_4 \cdot 2H_2O$$

The start of setting depends on the temperature of the gypsum decomposition; the higher the temperature, the slower the gypsum plaster setting. After having set, gypsum plaster begins to harden, i.e. it gradually gains strength and gypsum recrystallises. Higher strengths of gypsum plaster may be attained through its desiccation up to a temperature of 40°C. A considerable advantage of gypsum plaster is its capability of being completely hydrated and reaching final strengths in a relatively short time, i.e. up to about 3 days.

In spite of the fact that the product of gypsum plaster hydration is chemically identical to natural gypsum plaster, their internal structures are different. The volume mass of natural gypsum is 2300 kg/m³ and its compressive strength ranges from 40 to 45 MPa, while the volume mass of gypsum that was generated by the hydration of β-gypsum plaster is only about 500–1500 kg/m³ and the compressive strength ranges from 1 to 25 MPa. On the contrary, the compressive strength of gypsum formed by the hydration of α-gypsum plaster is higher, as much as 50 MPa.

The water contained in the hardened gypsum plaster is in three forms: chemically bonded water, capillary water inside the gypsum plaster in fine pores, and the water adsorbed on the surface. Hardened gypsum plaster shows hygroscopic properties. Therefore, it is not suitable to use gypsum plaster in the environment with a higher than 60% relative air moisture and in places where the gypsum plaster is in contact with water or soil moisture.

The quality of gypsum binders depends on the water/gypsum plaster ratio at mixing. This ratio is commonly within the range of 0.6–0.8. If more mixing water is present, it leaves pores in the gypsum plaster. Gypsum binders in the water/gypsum plaster ratio from 0.7 to 0.8 have pore contents within the range 47%–55% of the volume. The system of pores is continuous and makes it possible to receive and release moisture.

3.2 Deterioration of gypsum binder

To a certain extent, hardened gypsum plaster is, as already mentioned, soluble (241 mg in 100 g of water at 20°C) [12]. The damage of gypsum plaster binders may occur during longer contact with water.

The equilibrium moisture of gypsum plaster at 80% relative air moisture at temperature 20°C is 0.2 wt% [10]. The capillary condensation of water occurs in fine capillaries, and subsequently, the gypsum plaster binder is dissolved and may be transported by water to another place where it generates crusts afterwards.

In the presence of calcium ions or sulphates, the solubility of gypsum plaster decreases. On the contrary, if gypsum plaster comes into contact with different ions, its solubility increases. For example, a solution of NaCl with a concentration of 100 g/l increases the solubility of a binder by three times. Although the anion of sulphuric acid is identical, together with calcium sulphate this generates highly soluble additional compounds of the type $CaSO_4 \cdot H_2SO_4$ [11].

The suction capability of gypsum plaster is excellent; however, the sucked water is easily released depending on ambient conditions. This refers to physically bonded water that in pores is in the form of a so-called pore solution or is adsorbed on the surface in the form of water film.

Considerable changes of mechanical properties in relation to the product moisture represent the disadvantages of gypsum plaster. Strengths and the elastic modulus markedly drop with product moistening. At 1% moisture, the product strength drops by as much as 30%, and at 12.5% moisture by 50%. In view of these properties, gypsum plaster is suitable for use in an environment with relative moisture up to 60%. Gypsum plaster is considered frost-resistant if at least 20% of the pore volume is not filled with water.

The corrosive action of gypsum plaster on metals is one of its important chemical properties. The pore solution contains Ca^{2+} and SO_4^{2-} and its pH value is 5. A relatively fast degradation of steel profiles occurs by the action of H^+ ions. Products of steel corrosion appear in a short time on the surface of a gypsum plaster product and decrease the aesthetic value of the construction. Corrosion does not occur at relative air moisture up to 60% and product moisture up to 10%. Therefore, aluminium profiles or steel profiles provided with a protective coating are recommended for constructions combined with gypsum plaster binders.

Gypsum plaster binders are sensitive to temperatures higher than 40°C, at which water begins to be gradually released. Total decomposition to hemi-hydrate $CaSO_4 \cdot 1/2H_2O$ occurs as late as at temperatures above 110°C.

This property of gypsum plaster is utilised for the protection of constructions against fire. A gypsum plaster product releases bound hydrated water and loses strength at temperatures above 110°C. The released water generates a vapour layer that temporarily protects the construction against the effects of heat. The more massive the gypsum plaster lining is, the longer the structure may be protected. The thermal conductivity of gypsum plaster of common volume weight ranges in values from 0.25 to 0.55 W/m K.

3.3 Prevention measures

Hardened gypsum plaster is damaged by the action of water and by excessive mechanical load. It is known that mechanical properties of hardened gypsum plaster markedly depend on its moisture content. Desiccated gypsum plaster is two to three times stronger than moist gypsum plaster. For this reason, it is necessary to take into consideration the environment where gypsum plaster products are placed. This environment must not be moist.

In order to increase resistance to the effects of moisture, hydrophobic substances are applied on the surface of gypsum-based renders and plaster coatings. Especially, silicones that cause the magnification of the wetting angle over 90° are used, and thus they make the penetration of water into porous materials impossible.

Gypsum plaster is suitable for interiors; for exterior applications, gypsum plaster must be modified, using various kinds of polymer substances and also glass fibres. Colak [13] modified gypsum plaster by acrylic latex (methacrylic acid esters and styrene). In the gypsum plaster sample modified by 10 wt% of acrylic latex, the compressive strength was 24.6 MPa, while the reference sample showed 16.9 MPa. Colak also determined the strengths of samples that were immersed in water for a period of 7 days. After 7-day immersion in water, the unmodified gypsum with a water/gypsum ratio of 0.4 retains 50% of the original strength, whereas the latex-modified gypsum composites retain only 30% of the original strength. Water absorption is the same for the latex-modified gypsum and for the unmodified gypsum.

Bijen and van der Plas [14] modified gypsum plaster by polymer dispersion of polyphenylene sulphide and glass fibres. In the case of the former, a composition contains 13 wt % of glass fibres (wt % of total matrix) and 38.5 wt % of polymer dispersion (% wt in matrix). The durability of hardened modified gypsum was investigated in cycles of wetting and drying, heating and cooling, and UV-light loading. Results of 2400 hours or 400 cycles showed that modified gypsum appears not to be affected by this exposure. The glass fibre-modified sample was completely disintegrated very fast during this test. The tests exhibited that the polymer-modified fibre-reinforced gypsum has excellent weather ability characteristics. Gypsum plaster binders must come into contact with hydraulic binders containing aluminate components. In the presence of moisture, gypsum dissolves and reacts with its components together with the creation of ettringite and thaumasite [15]. Arikan and Sobolev [16] modified gypsum plaster using water-soluble polymer (modified methyl cellulose), air-entraining admixture (olefin sulphonate sodium salt), and superplasticiser (melamine formaldehyde sulphonate). They tracked the effects of admixtures on the consistency of fresh mixture, setting time, and compressive strength. They found that the superplasticiser and the modified methylcellulose significantly increase consistency, and modified methylcellulose improves mostly the compressive strength.

4 Bricks

Fired brick products are used in the building industry as masonry materials provided either with coatings or used as a non-coating masonry facing. Roof covering is another significant ceramic product used in the building industry. Brick-making has a long tradition, dating back to at least 3000 years. At the present time, the composition of ceramic materials and their damage in service may be studied in preserved structures.

The raw materials for the production of bricks are brick clays containing minerals of kaolinite, montmorillonite, halloysite, or illite type, and also accompanying compounds that may cause the damage of the finished product. These are especially calcite ($CaCO_3$), gypsum ($CaSO_4 \cdot 2H_2O$), magnesium sulphate, and vanadyl compounds. Non-plastic opening materials that prevent the formation of texture defects are admixed during the raw material treatment to a moulding mass. Small quantities of salts such as barium carbonate may be added to combat soluble salts. Barium carbonate reacts with soluble sulphates producing insoluble barium sulphate. The microstructure of the burnt ceramic body contains new compounds of the spinel type, an amorphous phase formed by the solidification of reaction products that owing to the effect of a high viscosity did not form crystals, the original unconverted minerals that are stable at the temperature of firing and also pores. The porosity of ceramic products is very important, and this is the root of many principles of damage.

In the past, besides fired bricks, mud bricks, which are even today the main masonry material in some countries, were used for building purposes. These bricks are made using the homogenisation of clays with sand and water [17]. Their service life in a dry environment is long, however, in long-term contact with water, these bricks are markedly damaged through elutriation.

4.1 Properties of bricks

Although the properties of bricks might be described by different parameters, the most important ones are as follows:

- porosity,
- strength,

- water absorption, and
- salt content in body.

Other properties, such as dimensions and appearance, are less significant, especially in bricks used under coating. The appearance is a very important property of facing bricks.

Porosity is the most important property of the brick body. The porous structure of the body depends, above all, on the temperature of firing and the content of melting oxides in the material. Apparent porosity is mostly capillary (pores are open and permeable), and determines the absorption capacity of the body. But on the other hand, the real porosity (PS) is calculated from the share of the volume density ρ_v and specific density ρ of the body according to the eqn (5):

$$PS = 1 - \frac{\rho_v}{\rho}. \tag{5}$$

The strength of the brick body and its permeability is also dependent on porosity. Solid brick strength ranges from 20 to 35 MPa, while the clinker (engineering brick), having a low porosity, reaches a compressive strength as high as 80 MPa.

The standard brick capillary radius dimension is from 0.006 to 6.0 μm. Sosim *et al.* [18] state that while at the temperature of firing 900°C, the total porosity ranges from 28% to 37% and the prevailing pore diameter is from 0.2 to 0.4 μm, at the firing temperature 1000°C, the total porosity is from 25% to 36% and most pores are of diameter from 0.4 to 0.8 μm, and at 1100°C, the total porosity is within the range of 7%–25% with the prevailing pore diameter from 8 to 1.6 μm. In practical applications, the firing temperature ranges from 950°C to 1050°C, whereas clinkers are burnt at temperatures above 1050°C. The microstructure of the common brick and clinker brick is given in Fig. 1.

The absorption capacity of the common brick ranges from 20% to 30%, while the absorption capacity of the clinker (engineering brick) is about 6%. Facing bricks are exposed to the direct action of atmospheric effects. The claims concerning the

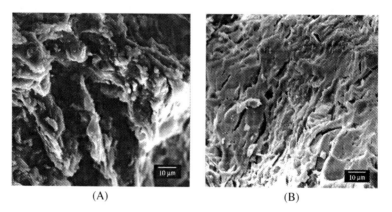

(A)   (B)

Figure 1: (A) Common brick; (B) clinker brick.

properties of these bricks are much higher. These types of bricks were used for construction that has survived several centuries, and yet some facing bricks are in good and undisturbed condition.

Bricks, especially facing bricks, are evaluated based on their resistance against the ambient environment. This evaluation includes the frost-resistance test and the determination of the content of water-soluble salts that may be contained in the brick body.

4.2 Effect of freezing

Products used for exterior applications (facing bricks and roof covering) must be frost-resistant in those climatic zones where the temperature in winter drops below the freezing point. The products of high-absorption capacities are not resistant to freezing and thawing, and decay after several cycles if saturated with water.

There exist several views of brick body damage caused by frost [19]. The capillary theory explains the damage due to the effect of change of the liquid water contained in the pores of a brick product into ice. Ice, compared to the water volume in the liquid phase, has a volume that is 9% higher; its crystallisation pressure causes the damage of the brick product structure. An illustration of brick masonry damage by freezing is given in Fig. 2.

Absorption capacity is, however, only an informative criterion for the evaluation of frost resistance. Frost damage also depends on the body microstructure, on the amount of water sucked by the porous system and further, on the velocity of the

Figure 2: Brick masonry damaged by frost.

lowering of temperature below the freezing point. Water adsorbed on the walls of capillary pores may remain in the liquid phase and at sub-zero temperatures.

When completely dry brick is water sprayed at sub-zero temperatures, it will adsorb water into the pore system in such a quantity as will exert vapour pressure equal to that of ice at the same temperature. Only such quantity of water that satisfies the requirements of the equilibrium may be adsorbed in the pores. Other shares of water will create ice before it may enter the pores. Consequently, dry bricks that are frozen cannot be water saturated and, therefore, the damage caused by the creation of ice in fine pores is not probable [20].

4.3 Effect of salts

Salts in brick masonry may come from three sources:

- the raw material,
- the subsoil, and
- the ambient atmosphere.

The most common sulphates in the raw material used are sulphates of calcium, magnesium, potassium, and sodium. When there is no horizontal insulation, underground water rises through the brick masonry. Such water contains dissolved salts that may be characterised by sodium, potassium, calcium, magnesium, and ammonium cations, and by anions such as sulphates, chlorides, and nitrates. The salts in the atmosphere that are in the air are in the form of aerosols sediment on the masonry surface. This is especially significant in coastal areas, but a source of sodium chloride may be the de-icing salt solution that is dispersed into the atmosphere by the wheels of vehicles. The salts on the masonry surface may also originate from the reaction of acid gases with water (CO_2, SO_2, and NO_x), together with the creation of acid that consequently reacts with the mortar components and forms soluble salts. Soluble ammonium salts are formed on structures that are near a source of ammonium ions. If soluble, the salts formed on the masonry surface create a solution under favourable humidity conditions and penetrate into the brick body. The penetration depth of a 10% solution of sodium sulphate into the brick and its crystallisation are shown in Fig. 3.

When in solution, salts are harmless for the brick masonry. The concentration of these solutions in brick pores increases with the quantity of evaporated water. On the contrary, if capillary condensation occurs, the condensation in pores decreases. Salts may damage bricks and the total brick masonry in the following ways:

- By the crystallisation of salts close to the surface or in the brick body surface, Fig. 3; the crystallisation pressure causes the development of pressure on the pores walls. This pressure may be so high that it disrupts the brick surface.
- By the recrystallisation of salts where hydrates of a higher number of water molecules usually are formed.
- Sulphates may be transported from bricks into the bedding or pointing mortar and, because these mortars contain silicate and aluminate phases, a reaction

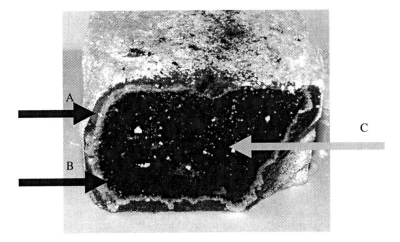

Figure 3: Penetration depth of sodium sulphate solution into brick. (A) Brick surface; (B) salt crystallisation; (C) intact brick body.

proceeds, and either ettringite $3CaO \cdot Al_2O_3 \cdot 3CaSO_4 \cdot 31H_2O$ or thaumasite $CaSiO_3 \cdot CaSO_4 \cdot CaCO_3 \cdot 15H_2O$ are created [21]. The molecules of both the compounds have big molar volume causing the development of considerable pressure on the surrounding mortar, and the mortar begins to decay.

The greatest damage of bricks leading finally to their weathering is caused by the repeated creation of crystals. During the period of time when there is a high content of water in the masonry, only insoluble or less soluble salts crystallise. If the high content of humidity is exchanged by the masonry desiccation, the concentration of salts in water contained in the masonry increases, and even the highly soluble salts crystallise. In some areas, the rotation of the periods of drought and rains is regular, and the crystallisation of salts always occurs at the beginning of the period of drought [22].

4.4 Pollution prevention measures

Based on the information dealing with masonry damage presented earlier, it may be concluded that water and its transportation may be the cause of the brick masonry damage; however, this is not the only cause. Measures that may be taken to prevent masonry damage by efflorescence, the crystallisation on the surface of bricks, and by the action of frost may be divided into three groups:

- *Reduction of salt content and humidity in the masonry:* This may be carried out by means of additional horizontal insulation either by injecting that will form a waterproof band in the masonry or by inserting watertight face insulation. In this way, the capillary lift of ground water into the masonry and the transportation of water-soluble salts are prevented. The reduction of the salt content in the brick body's porous system reduces the risk of damaging the bricks by freezing that may result in their decay.

- *Reduction of salt content in the brick body:* This may be accomplished by the proper selection of raw materials and by adding substances which together with soluble salts create insoluble compounds that cannot participate in the transportation of ions.
- *Providing the masonry surface with water repellent coatings:* The most important advantages of water repellent treatment are [23] stopping direct water leakage, preventing frost damage, reducing salt efflorescence, chemical degradation, and biological attack.

The majority of water repellents are based on silicone materials. These compounds contain a very strong bond Si-O-Si that is stable and resistant to water action, the action of chemicals and also to UV radiation. With respect to its long-term stability and good adhesion, water-repellent coatings based on silicones have a long service life.

5 Natural stone

Stone is one of the oldest building materials. Its resistance to mechanical damage and to the action of the atmosphere is high enough. Nevertheless, we know that in the course of time even stone is subject to irreversible changes leading to its damage, and sometimes to its total disintegration. This process may be retarded by suitable care.

The causes of stone damage may be divided into three groups. *Physical damage* includes the effects of temperature and humidity changes including freezing, crystallisation of salts and the effects of water. *Chemical damage* is caused by the reaction of corrosive substances from the ambient environment with compounds forming relevant rock. In most cases, this damage results in the creation of soluble compounds, and their washing out from stone that causes the increase of its porosity and finally, its decay. *Living organisms* may cause mechanical damage, e.g. by plant roots that during growth, cause the mechanic damage of building materials. Stone is also subject to damage by products of metabolic processes of living organisms, e.g. by excrements of animals or by products of micro-organisms.

5.1 Physical influences

Rocks are heterogeneous materials composed of constituents of different thermal expansions. This expansion may change depending on the crystal form in which the mineral forming the rock occurs. Sometimes, the thermal conductivity is different in some respects. Rocks have low thermal conductivity, which is the cause of the origin of the thermal gradient between the stone surface layer and its centre when the ambient temperature changes. Under high temperatures during fire, the difference between the stone surface layer and centre is still more distinctive. The abovementioned phenomena result in the creation of fine cracks that deteriorate the mechanical properties of stone and enlarge the surface on which a chemical reaction with the environmental constituents and further notable damage of stone may occur.

Stone, like any other material, has a certain equilibrium humidity that is dependent on the temperature and humidity of the ambient environment. Based on the stone's pore structure, its mass absorbs liquid rainwater, dew, water from ice and hoarfrost, and water brought by wind from large bodies of water in the form of aerosol. In ground floor parts of buildings, water from subsoil creeping up to stone depending on the size and shape of its pores may play an important role. Liquid water also gets into stone when the stone is cleaned and restored. Then, the transport inside stone proceeds in two directions: horizontal and vertical. The horizontal motion of water may be described in a simplified way according to Fick's law by the equation:

$$x = A\sqrt{t}, \tag{6}$$

where x is the depth of water penetration into stone in time t and A is a constant that may be expressed by the relation:

$$A = \sqrt{\frac{\sigma r}{2\eta}} \tag{7}$$

where σ is the surface tension of water, η is water viscosity, and r is the radius of the pores. It follows from the relation that the penetration of water into stone depends on the size of the pores. The larger the pores, the greater the penetration. The equilibrium between the motion of water upward and the gravitation is established in the horizontal direction according to eqn (1). Both relations describe the motion of water in stone with certain limitations, since the pores are not formed by straight tubes and the water is not clean and contains dissolved salts.

Water, which is freely movable in pores, increases its volume by 9% in transitioning into ice, which results in stone damage by the crystallisation pressure. Water that is adsorbed on pore walls does not move. The arrangement of molecules of this water is regular and may be compared with crystals, and occurs at optimum temperatures from 20°C to 30°C. Due to the arrangement of the molecules of water, the formation of pressures whose dimensions are comparable with the crystallisation pressures of ice is assumed to proceed in small pores.

There also exists the mechanical action of water. Flowing water is a cause of building stone damage through abrasion where the water flow brings fine particles of solid materials that damage stone mechanically; another mechanical damage is caused by erosion or cavitation. This type of damage may be encountered in water structures, especially in stone bridges.

The action of wind that damages stone mechanically may also be included in physical effects. The mechanical stress stems from impacts of wind; however, local failures and stone abrasion are especially caused by fine solid particles carried along with it. Rainwater and snow on the windward side are driven by wind to the stone surface; on the other hand, in a dry period, wind contributes to its desiccation.

5.2 Chemical effects

Building stone is affected by chemical constituents of the atmosphere or by salts dissolved in water in subsoil; these salts are either picked up by the stone or are in contact with the stone in the form of soil moisture. These substances react with the rock components and create new, more soluble compounds or, vice versa, create insoluble compounds of higher molar volume.

Besides the basic components (nitrogen, oxygen, noble gases, and carbon dioxide), air also contains water vapour and gaseous or solid products of industrial origin, although there exist very strict limits for the pollution of the atmosphere at the present time. The presence of sulphur dioxide and nitrogen oxides represents the most serious hazard for stone damage. Carbon dioxide is sometimes rather neglected, ignoring the fact that Nature is not able anymore to overcome the problems of its increased concentrations resulting from the growth of traffic, industrial production, and power engineering, and that the average content of carbon dioxide in the atmosphere has increased. Acid constituents of the atmosphere together with water form dilute solutions of inorganic acids, and rainwater may then drop the pH value below 4. Sulphurous, sulphuric and nitric acids, and also carbon dioxide (either dissolved in the molecular form or as carbonic acid), are most frequently contained in the atmosphere in the form of aerosols. These acids particularly react with carbonate rocks such as limestone and dolomites; however, these acids also attack the rock binders based on calcium carbonate, $CaCO_3$. In the reaction of sulphuric acid with $CaCO_3$ (solubility 1.4 mg in 100 g of water at 20°C [12]), the resulting product is gypsum, $CaSO_4 \cdot 2H_2O$ (solubility 241 mg in 100 g of water at 25°C [12]), which is significantly more soluble than the original calcium carbonate.

Soluble calcium bi-carbonate, $Ca(HCO_3)_2$, forms from the reaction of carbon dioxide dissolved in water with $CaCO_3$. It is more soluble (150 mg in 100 g of water at 20°C) than calcium carbonate. Calcium bicarbonate exists in solution only, and under suitable conditions it reverses back to $CaCO_3$. So, it is evident that this reaction is closely connected with the presence of liquid water [24].

Acid hydrolysis of other minerals, e.g. feldspars or iron compounds, also occurs in the acid environment. This hydrolysis results in the formation of new compounds with properties different from those of the original compounds.

Salts soluble in water are sometimes present in the rock itself but, in most cases, these are transported into the stone. As far as cations are concerned, building stone most frequently contains calcium, potassium, and ammonium, and as to the anions, sulphates, chlorides, carbonates, and nitrates are present in stone.

Sulphur oxides from the atmosphere, sulphates from the subsoil, or dust containing sulphates and depositing on the stone surface are the usual sources of sulphate in stone. The source of chlorides is different owing to the area where the structure is situated. Salts used for the winter maintenance of roads and pavements are common sources. The most significant source of chlorides and also sulphates in coastal areas is the aerosol of seawater in the air. Nitrates, especially ammonium nitrate, NH_4NO_3, usually have their origin in the pollution of the structure surroundings by faeces, in synthetic fertilisers, or in the accumulation of the excrements and urine of animals directly on the stone.

As explained in Section 2.2, crystallisation or recrystallisation of salts occurs. The penetration of highly soluble salts into stone is deep, and these solid salts precipitate during the evaporation of a substantial portion of the water from the pore solution. Salt hydrates may lose water during the reduction of the relative humidity of air. When the relative humidity increases, these hydrates form crystalline hydrates. Crystallisation and recrystallisation of salts are connected with the change of the molar volume that demonstrates itself in pressures on the walls of pores in which the crystals were formed. These pressures may exceed the stone strength and, in such cases, the stone decays. This phenomenon occurs if salt crystallises in the stone layers close under the surface. Such a layer is usually about 1-mm thick and after its destruction this phenomenon advances into another subsurface layer. Continuous disintegration of such layers proceeds through this gradual process.

The dilution of solutions that penetrate into stone in the horizontal direction leads to a deeper penetration. In contrast, the more concentrated solutions of higher density remain in the surface layers where water evaporates more easily and, thus the crystallisation of salts is easier. Water also evaporates in the stone evaporation zone and the salts form crystals on the stone surface if evaporation prevails over the supply of the capillary action water. The places forming wetting boundaries, i.e. where the evaporation prevails over the supply of the capillary action water and where the solutions are concentrated and subsequently crystallise, are subject to the largest damage.

5.3 Biological effects

Bacteria, algae, fungi, lichens, higher plants, and animals may damage stone both mechanically and chemically. The mechanical action of plants is limited to the root system. It has been proved that the growth parenchyma placed in the end of the weakest root fibre is capable of generating a pressure from 15 to 35 MPa.

The chemical action of living organisms on stone is very diverse. Lichens and fungi produce organic acids that are especially damaging to carbonate rocks. Higher plants may also damage stone structures chemically by the action of roots diluting humic acids that attack carbonate rocks and disintegrate them. Serious stone damage may be caused by flocks of birds. In conurbations, these are mainly pigeons whose excrements contain components such as compounds of phosphorus, nitrogen, and sulphur, which are the nutrient medium for bacteria. The ground floor parts of structures polluted by dog's urine are also a suitable environment for the action of micro-organisms. When the humidity content is approximately 5 wt %, various kinds of bacteria living from the compounds and impurities in stone and producing compounds called metabolites increase in number and grow in the stone surface layer. In this layer, sulphur bacteria transform sulphur oxides into sulphuric acid, and nitrification bacteria transform ammonia into nitric acid. If generated, both acids participate to a large extent in the damage of stone components.

Bacteria also participate in the disintegration of Si-O-Al bonds in aluminosilicates by penetrating into the structure of minerals, where they cause the bonding of a proton to oxygen resulting in the formation of Si-OH-Al. By attacking the bonds that were originally very consistent, simple compounds such as amorphous $Al(OH)_3$ and $Si(OH)_4$ are produced. The extent of the microbiological degradation of these minerals rises with the increasing content of Al in the compound [25].

Stone exposed to the ambient atmosphere is covered with a thin layer of salt, dust and dead bodies of micro-organisms. This relatively thin, compact, usually black layer, called the crust or sometimes also patina, is less harmful for stone of low porosity and resistant to compounds based on silicates. The crust contains some organic compounds, above all of paraffin type, which lends to the water repulsion of the crust surface. The crust on rocks containing carbonates is usually formed by gypsum produced by the reaction of sulphur compounds with $CaCO_3$. The porosity of the stone surface increases, while the porosity of the crust is low. Owing to its dark colour, the crust is perceived as soiling that is dependent on many factors, especially of the type of stone, type of exposure, type of climate and on the covering against rain and snow [26]. An investigation of the sandstone damage on the adornment of St. Peter and Paul's Cathedral in Brno is presented in Section 6.

5.4 Types of building stone and its damage

Rocks used as building stone may be divided into three groups.

5.4.1 Volcanic rocks

Volcanic rocks originate by the cooling of magma. Their main components are silicates with a significant content of aluminium. Volcanic rocks show excellent strength and hardness. Their porosity is low and they are resistant to weathering. Granite, which is most frequently utilised in the building industry, is a typical representative.

5.4.2 Granite

Granite is a granular rock of different grain sizes. Its colour is usually light grey, sometimes blue-grey, yellowish, pinkish, or also red. It is composed of potassium and calcium feldspar grains, quartz and dark silicates, and mica. Granite is used in the building industry as building stone, paving stones, steps, and as a material for sculpture. Due to its composition, granite is very resistant to the action of corrosive compounds from the surroundings; its porosity is very low, which also results in a low absorption capacity. After 35 freezing and thawing cycles in a solution medium of NaCl and Na_2SO_4, granite was not damaged. The negligible weight loss of granite was caused by the loss of mica grains [27].

5.4.3 Sedimentary rocks

Sedimentary rocks originate by the sedimentation of the products of weathering of primary rocks. These often contain solid frames or skeletons of animals, fragments of rocks, clayey minerals, and chemically precipitated substances. These rocks contain carbonates, silicates, oxides, hydroxides, and sulphates. In most cases, these rocks are less resistant to the action of weather, and are shaped more easily than volcanic rocks. The most frequently used rocks for structural purposes are limestone, sandstone, and arenaceous marl.

Limestone is usually composed of calcium carbonate in the form of calcite, with frequent admixtures of clay and dolomite. In the building industry, it is used as gravel and building or decorative stone.

Sandstone originated by the sedimentation of fragments of volcanic, metamorphic or sedimentary rocks. The grains are mostly of quartz and depending on their size, sandstone is classified as fine-grained, medium-grained, or coarse-grained. The grains are bonded with a binder that may be based on silicate, kaolinite, aluminate, or ferrite. Silicate grains in sandstone are highly resistant to the action of environmental effects and, therefore, the binder quality is the decisive criterion of sandstone quality. When sandstone has the calcite binder ($CaCO_3$), which is decomposed by corrosive components from the atmosphere, is the most susceptible to damage. The porosity of some kinds of sandstone is very high, which results in a higher absorption capacity. These kinds of sandstone have lower strength and resistance to weathering compared with sandstone of low porosity.

Arenaceous marl is the name for rock that contains quartz grains, consolidated clay, mica, and other components. Calcium carbonate is the basic binding constituent. The high absorption capacity of arenaceous marl results from its significant porosity. It also has a high-equilibrium humidity, which manifests itself as stone cracking below the freezing point. The excessive drying of arenaceous marl leads to the formation of harmful stresses and to the mechanical damage of the stone. Owing to the sedimentary characteristic of the rock, its character is anisotropic.

Arenaceous marl containing calcium carbonate is not very resistant to the action of acid gases from the atmosphere that decompose $CaCO_3$ together with the creation of more soluble compounds (sulphates, nitrates, chlorides). The chemical deterioration results in the surface layer dropping off and in the uncovering of a new surface.

Limestone, like arenaceous marl, is also not very resistant to acid gases from the atmosphere. If its porosity is high (shell limestone and travertine), corrosive substances penetrate into the depth and the disintegration proceeds faster than in low-porous limestone that is more resistant. The damage of sandstone depends on the composition of the binder connecting the quartz grains and on porosity [28]. The carbonate binder is less resistant than the quartz or aluminosilicate binder.

5.4.4 Metamorphic rocks

Metamorphic rocks originate by the recrystallisation of sedimentary or volcanic rocks under high pressures and temperatures. Through these processes, a more stable form of crystals of a similar chemical composition compared with the original rocks was created. Marbles that originated by the recrystallisation of limestone rank among the most significant metamorphic rocks. Marbles are dense and compact, and their porosity and absorption capacity are low. Colouring admixtures create the colour drawings in marbles; pure calcium carbonate is white. Coarse-grained and dolomite marbles are more resistant to the action of the surrounding environment than fine-grained and calcium marbles. The damage of polished marbles manifests itself in the loss of lustre and change of colour.

5.5 Pollution prevention measures

As described above, many changes of properties occur during stone weathering, especially the change of porosity, absorption capacity, colour, structure, mechanical properties, and sometimes even the change of the surface shape, or also the loss of cohesion. Detailed research of the existing state of stone is necessary before any intervention of the restorer or conservator can proceed. Based on such research, it is decided what measures will be taken.

In most cases of stone damage, the first phase consists of stone surface cleaning that may be carried out chemically (by washing with water or with a detergent solution) or mechanically (by removing the impurities in dry places with a brush or using a cleaning paste).

For the desalting of stone, it is possible to use various poultices, such as cellulose pulp, a clay poultice or coating of lime mortar. There is good experience with using poultices for stone desalting, although it should be mentioned that the desalting is partial only.

Electrochemical extraction has also been used for stone desalination. This extraction was successfully used for removing chlorides from concrete [29]. The removal of salts from stone did not lead to definite conclusions and, therefore, this method is not recommended for stone desalination [30].

Biotechnological methods that are currently being developed are interesting [31]. The use of denitrification bacteria may transform nitrates to gaseous nitrogen in the course of 6–12 months. Cations remain in the material and combine with HCO_3^- generated from CO_2 close to stone.

Utilisation of crystallisation inhibitors offers the possibility of controlling (inhibiting) salt crystallisation, thereby inducing salt growth as efflorescence and minimising damage to the porous substratum [32].

Damaged stone is consolidated with inorganic or organic consolidation agents. The applied substances must be stable, resistant to light, oxidation, moisture, must be chemically inert, and resistant to the action of fungi and bacteria. Of the consolidation agents, the solution of calcium hydroxide is often used. Sometimes, it is replaced by barium hydroxide. Organic consolidation agents based on wax, silicones, epoxides, methyl acrylates, and polyurethanes are used for stone consolidation much more frequently. Both the consolidation of stone and other restoring measures must be taken individually to avoid causing irreversible damage.

Surface hydrophobisation is an effective measure against stone damage. Hydrophobic substances such as oil and wax used in the past are replaced today by silicones, silanolates, and esters of silica sol.

6 Deterioration of the sandstone adornment of the St. Peter and Paul's Cathedral in Brno

The St. Peter and Paul's Cathedral in Brno is the great architectural heritage which is shown in Fig. 4. The destructive influence of gypsum on silicate sandstones on stone adornments of the facing leaf was studied. It was discovered that gypsum could be for silicate sandstones one of the particular destructive factors.

The type of the sandstone used on the cathedral is relatively common, and it was often used in Bohemia and Moravia. The gypsum coat on the sandstone was noted also on other monuments made from a similar type of rock, e.g. on the statuary adornment of the Charles Bridge in Prague.

After the neo-Gothic reconstruction of the whole cathedral, the first of more extensive repairs were conducted from 1976 to 1978. Failed and damaged elements were re-attached, completed, or fully replaced with elements of artificial sandstone. The reconstructed elements had been produced using silicone rubber moulds. Diluted polymer dispersion (PVAC) was used for stone conservation in some places.

Figure 4: The St. Peter and Paul's Cathedral.

The deterioration of the adornment parts over the next 18 years was of high degree, which negatively affected the aesthetic appearance of the cathedral. Some elements of the adornment collapsed and many others were loose and at risk of falling. The worst situation was on the western front, where it was necessary to immediately remove the loose elements that were endangered to falling. It has been discovered that sandstone elements, which were made from the 'lighter rock', positioned below the elements that are shown with an indicator in Fig. 5, were the most damaged parts. It has turned out that this 'lighter rock', which was reported in archives as 'the light sandstone', is in fact limestone. In such damaged places the deterioration of the rock appeared particularly by sanding and by the gradual loss of cohesion in depth, or by the formation of thin cracks, respectively. In some elements the stability was at risk due to the decay. In comparison with limestone elements, sandstone surfaces were darker.

Damage of stone adornments was characterised by the following:

- sandstone had a dark layer of deposits;
- limestone elements were decayed in most cases due to the local formation of a firm crust;
- limestone pieces were usually less deteriorated than the sandstones placed below them; and

Figure 5: Decayed sandstone pinnacle stem under the limestone pinnacle.

Figure 6: Gypsum crystals in the pores of sandstone.

Table 4: Principal characteristics of stones.

Stone	Bulk density (kg/m³)	Specific density (kg/m³)	Water absorption (wt %)	Porosity (wt %)
Biodetrital limestone – light rock	1860	2530	11.6	26.5
Silicate sandstone	2130	2610	9.2	22.4

- below the repaired materials from the 1970s the stone layers were heavily deteriorated.

The principal characteristics of both kinds of stones are shown in Table 4.

The anions of sulphates, nitrates, and chlorides were determined in the samples of sandstone and limestone that were taken from the places with different extents of decay. The highest content of sulphates was found in deteriorated places. The presence of calcium sulphate and gypsum were determined by optical microscopy. Figure 6 exhibits a thin-section of sandstone in transmitted light, crossed polars, magnification 100×. The filling of the pore space by gypsum crystals is visible in the centre.

The chemical damage of the sandstone itself cannot serve as the source of gypsum, as the sandstone contains no compounds (calcite) convertible to calcium sulphate. In this case, calcium sulphate is transported into sandstone from the limestone following the reaction of calcium carbonate with sulphur oxides from the atmosphere in the presence of water. A small contribution might theoretically also come from the accumulation of dust or from the conversion of airborne carbonate particles deposited on the sandstone surface. The particularly high amounts of sulphate found in the neighbourhood of the pointing mortar may also come from these materials. With respect to the obvious differences in gypsum contents between elements underlying limestone and other materials, and regarding the calcium sulphate concentration on pinnacles, it is logical to regard the adornment elements made of limestone as the main source of sulphates.

The most common and extensive decay of the cathedral's outer stone adornment occurred on the sandstone positioned below limestone elements. In all the investigated places the calcium sulphate content was several times higher in the deteriorated sandstone underlying the limestone than in unaffected places of the adornment made from sandstone itself.

The compaction by gypsum has a considerable influence on some sandstone properties with respect to water migration: the rock surface is sealed by gypsum and hence the water transport is hindered. As the sandstone type used contains no carbonates and cannot serve as the source of gypsum, products of chemical deterioration of the adjacent limestone elements are considered as the essential source of gypsum in sandstone. As the limestone adornment elements contain similar amounts of gypsum as the underlying sandstone, the former exhibit as frequently as the latter significant damage phenomena. Both litho types have a similar absorption coefficient for water, similar gas permeability and sorption isotherms.

The influence of gypsum on the sandstone deterioration is principally apparent, and one can assume possible mechanisms of action as follows:

• pressure of crystallisation related to gypsum formation;
• changes of physical–mechanical properties of the sandstone following its saturation with gypsum, such as water exchange between the stone and the environment, E-modulus, hygric dilatation; and
• changes in the pore size distribution can be the reason for the reduced resistance of some external influences, e.g. frost.

In the decayed sandstone, gypsum was identified as the only form of calcium sulphate; no other types such as anhydrite or bassanite were detected.

Acknowledgements

The author would like to thank Dr Karol Bayer from the Institute of Restoration and Conservation Techniques, Litomyšl, Czech Republic, for kindly providing the results of the investigation presented in Section 6.

Nomenclature

A	constant
g	acceleration due to gravity
h	height of capillary action
m	mass of water vapour
M	molar mass

M_H hydraulic modulus
pH   negative logarithm of hydrogen ion concentration
PS   porosity of brick body
r capillary or pore radius
R_g universal gas constant
t time
T temperature
V volume
x depth of water penetration
η viscosity
φ relative humidity of air
Φ absolute humidity of air
Φ_{nas} humidity of the water vapour-saturated air
λ thermal conductivity
θ angle of wetting
ρ specific density
ρ_l pressure in the liquid phase
ρ_v volume density
σ surface tension

References

[1] Černý, R. & Rovnaníková, P., *Transport Processes in Concrete*, Spon Press: London and New York, pp. 33–45, 2002.

[2] Rovnaníková, P., *Coatings: Chemical and Technological Properties*, STOP: Prague, pp. 14–34, 2003 (in Czech).

[3] Greenwood, N.N. & Earnshaw, A., *Chemistry of the Elements*, Informatorium: Prague, p. 857, 1993 (in Czech).

[4] Heidingsfeld, V., Physical and chemical corrosion of building materials. *Proceedings of the Seminary on Water – Enemy of Architectural Heritage*. STOP: Prague, pp. 9–12, 1997 (in Czech).

[5] Davidovits, J., Ancient and modern concretes: what is the real difference? *Concrete International*, **9(12)**, pp. 23–28, 1987.

[6] Sabboni, C. *et al.*, Atmospheric deterioration of ancient and modern hydraulic mortars. *Atmospheric Environments*, **35**, pp. 539–548, 2001.

[7] Martínez-Ramírez, S., Puertas, F., Blanco-Varela, M.T. & Thompson, G.E., Studies on degradation of lime mortars in atmospheric simulation chamber. *Cement and Concrete Research*, **27**, pp. 777–784, 1997.

[8] Martínez-Ramírez, S., Puertas, F., Blanco-Varela, M.T. & Thompson, G.E., Effect of dry deposition of pollutants on the degradation of lime mortars with sepiolite. *Cement and Concrete Research*, **28**, pp. 125–133, 1998.

[9] Carlsson, T., Relationship between the air entrainment system and frost durability in hardened mortar. *Durability and Building Materials and Components 7*, Vol. 1., ed. C. Sjöström, E & FN Spon: London, pp. 562–570, 1996.

[10] Schulze, W, Tischer, W., Ettel, W. & Lach, V., *Non-Cement Mortars and Concrete*, SNTL: Prague, p. 65, 1990 (in Czech).

[11] Remy, H., *Inorganic Chemistry*, SNTL: Prague, p. 295, 1961 (in Czech).

[12] Lide, D.R. (ed.), *Handbook of Chemistry and Physics*, CRC Press: London, pp. 4–49, 1994.

[13] Colak, A., Characteristic of acrylic latex-modified and partially epoxy-impregnated gypsum. *Cement and Concrete Research*, **31**, pp. 1539–1547, 2001.

[14] Bijen, J. & van der Plas, C., Polymer-modified glass fibre reinforced gypsum. *Materials and Structures*, **25**, pp. 107–114, 1992.

[15] Skalny. J., Marchand, J. & Odler, I., *Sulfate Attack on Concrete*, Spon Press: London and New York, p. 46, 2002.

[16] Arikan, M. & Sobolev, K., The optimization of gypsum-based composite material. *Cement and Concrete Research*, **32**, pp. 1725–1728, 2001.

[17] Ren, K.B. & Kagi, D.A., Upgrading the durability of mud bricks by impregnation. *Building and Environment*, **30(3)**, pp. 433–440, 1995.

[18] Sosim, S., Zsembery, S. & Ferguson, J. A., A study of pore size distribution in fired clay bricks in relation to salt attack resistance. *Proceedings of the 7th Int. Brick Masonry Conference*, Melbourne, 1985.

[19] Hall, Ch. & Hoff, W.D., *Water Transport in Brick, Stone and Concrete*, Spon Press: London and New York, pp. 247–257, 2002.

[20] Zembery, S. & Macintosh, J.A., Survey of the frost resistance of clay brickwork in the Australian Alps, Research paper 6, CBPI, 2002.

[21] Halliwell, M.A. & Crammond, N.J., Deterioration of brickwork retaining walls as a result of thaumasite formation. *Durability of Building Materials and Components 7*, Vol. 1., ed. C. Sjöström, E & FN Spon: London, pp. 235–244, 1996.

[22] Kuchitsu, N., Ishizaki, T. & Nishiura, T., Salt weathering of the brick monuments in Ayutthaya, Thailand. *Engineering Geology*, **55**, pp. 91–99, 1999.

[23] Sandin, K., Masonry water repellent. *Durability of Building Materials and Components 7*, Vol. 1., ed. C. Sjöström, E & FN Spon: London, pp. 553–561, 1996.

[24] Zelinger, J. et al., *Chemistry for Conservators and Restorers*, Academia: Praha, pp. 192–235, 1987 (in Czech).

[25] Wasserbauer, R., *Biological Deterioration of buildings*, ABF: Prague, pp. 92–94, 2000 (in Czech).

[26] Grossi, C.M. et al., Soiling of building stones in urban environments. *Building and Environment*, **38**, pp. 147–159, 2003.

[27] Wessman, L.M., Deterioration of natural stone by freezing and thawing in salt solution. *Durability of Building Materials and Components*, Vol. 1., ed. C. Sjöström. E & FN Spon: London, pp. 342–351, 1996.

[28] Uchida, E. et al., Deterioration of stone materials in the Angkor monuments, Cambodia. *Engineering Geology*, **55**, pp. 101–112, 1999.

[29]   Mietz, J., *Electrochemical Rehabilitation Methods for Reinforced Concrete Structures*, A State of the Art Report, The Institute of Materials: London, 1998.

[30]   Vergés-Belmin, V., Desalination of porous materials. Workshop *SALTeXPERT*, ARCCHIP and Getty Conservation Institute, Prague, 2002, CD.

[31]   Wilimzig, M., Desalting of nitrate by denitrification. Workshop *SALTeXPERT*, ARCCHIP and Getty Conservation Institute, Prague, 2002, CD.

[32]   Rodrigues-Navaro, C., Fernandes, L.L. & Sebastian, E., New developments for preventing salt damage to porous ornamental materials through the use of crystallization inhibitors. Workshop *SALTeXPERT*, ARCCHIP and Getty Conservation Institute, Prague, 2002, CD.

CHAPTER 8

Environmental deterioration of building materials

R. Drochytka & V. Petránek
*Department of Technology of Building Materials and Components,
Faculty of Civil Engineering, Brno University of Technology, Brno,
Czech Republic.*

1 Introduction

By the durability of individual building materials and of the structures built of them, it is the time of usability under preservation of their original properties is understood. The words corrosion and deterioration are primarily derived from this term. Corrosion is applicable to metals, reinforcement and polymers. For silicate building materials such as concrete, plasters, mortars, we use the term deterioration. Deterioration of concrete is a gradual and irreversible permanent erosion of its basic properties leading to its disintegration. The study of concrete and building material deterioration and durability and the knowledge of these properties form the fundamental basis for the evaluation of structures. The durability of structures is influenced by many factors. The general defects and failures of building structures can be divided into the following (see Fig. 1):

- deterioration by the environment,
- defects and failures in design and in completion of structures,
- unsuitable design,
- bad quality of material,
- bad execution, and
- failures caused by inappropriate utilisation.

Defects and failures of buildings can be eliminated by proper control of products and execution of construction, and inappropriate utilisation of materials can also be eliminated. Generally, we cannot influence the environment much.

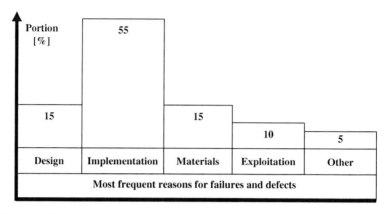

Figure 1: The graph shows the causes of failures in individual constructions.

For this reason, it is necessary to select suitable materials for structures with the knowledge of their deterioration processes in a given environment. Different types of environment have been described in the earlier chapters of this book. This chapter provides practical examples of construction deterioration. Buildings were mostly damaged by the atmosphere as a result of CO_2 effect (carbonation), SO_2 effect (sulphation) and by different forms of water effect. The last cause manifests itself during condensation and frost effect and even afterwards as water supply for the necessary chemical reactions. Effects of groundwater in the form of dilatation are mentioned as examples of corrosive liquid media. Finally, biological effects and the influence of stray currents are represented which, besides their corrosive activity on the reinforcement of reinforced concrete, can negatively act in different forms of corrosion, for instance, in reaction with sulphates.

2 Damage to residential and office buildings

The amount of influence of environmental deterioration depends mainly on the kind of building material used. The influence of the building structure is not so significant. The kinds of building materials were significantly determined in the past by local availability and transport limitations. This fact has an influence even at the present time when the selection of building materials is influenced by tradition, especially with regard to residential buildings built by individuals.

Building materials for building of family houses consist mostly of wood, ceramic and bricks materials, concrete (blocks, components), cellular concrete, blocks from alternative materials – sawdust concrete, stone in some regions and other alternative materials, as the case may be. The same types of structures and materials are currently used in such cases for large residential and office buildings too.

This concerns mostly skeleton building systems with a load-bearing structure of reinforced concrete or steel. The internal division and jacketing is made from prefabricated cellular panels, concrete or ceramic products.

2.1 Concrete and reinforced concrete structures

Concrete, which is a material with alkaline reaction, can be attacked by many corrosive factors. The common corrosive medium for all reinforced concrete structures of residential and office buildings are corrosive gases of acidic character. These corrosive gases are contained mainly in pollutants. The corrosive effects of gases from the atmosphere are significant, in particular, if gases can penetrate into concrete.

Among corrosive gases, carbon dioxide CO_2, sulphur dioxide SO_2, sulphur trioxide SO_3 and the nitrogen oxides are important. The CO_2 concentration in the atmosphere is on the whole stable; it is about 60 mg in $1 \, m^3$ of air. A great increase of concentration can be observed in CO_2 sources, for instance, in large deposits of natural springs. The normal SO_2 concentration in the natural environment does not exceed 0.01 mg in $1 \, m^3$ of air. An increased SO_2 concentration is identified in large industrial centres. The nitrogen oxides are mostly secondary products of motor traffic, but we can find them in great quantities in atmosphere polluted by chemical industries, and they are mostly not the decisive corrosive factor.

Water vapour from the atmosphere, much like carbon dioxide and other gases, penetrates capillary, porous materials by diffusion. The moisture penetrates into building structures not only in the gaseous state but also in the liquid state (capillary elevation, leaking in, etc.). Diffusion cannot take place in pores filled with water. The course of diffusion determines if corrosive chemical reactions will take place at all and, if so, under what conditions. Diffusion thus has an influence on the results of corrosive reactions.

2.1.1 Carbonation of concrete structures

Carbon dioxide in contact with cement that is formed by basic hydration binder products starts a neutralisation reaction – carbonation. The main products of this reaction are different carbonates.

The carbonates formed by carbonation generate new formations of insoluble $CaCO_3$ sediment in pores and capillary tubes which are filled by them step by step. This process decreases the possibility of new CO_2 penetration. In dense concrete it causes an external decrease of the concrete's permeability. This factor is very important as it explains why the perfectly dense concrete is resistant against the harmful effect of carbonation and why the steel reinforcement in it remains protected against corrosion for many decades.

The most common kinds of concrete, especially those of lower and medium strength class, have a higher porosity that cannot be substantially lowered by the new carbonate formations and thus results in lower permeability of concrete to gases and vapours. Carbonation progresses into still greater depth, attacks the adjoining cement paste and, due to the decrease of pH, causes a very strong corrosive reaction of the reinforcement. The effects of carbonation on reinforced concrete structures can be expressed by two consequences (see Fig. 2).

When concrete is saturated with water, the micro pores of the concrete are practically completely filled with water and the water film prevents deeper gas penetration into the internal parts of the concrete, even if the gases are dissolved in

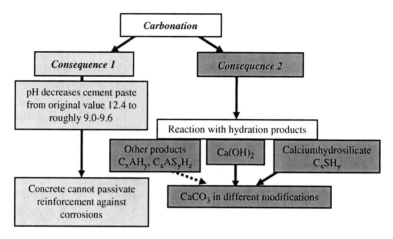

Figure 2: Scheme shows two different consequences of carbonation. Nomenclature: C – CaO, S – SiO$_2$, A – Al$_2$O$_3$, H – H$_2$O [1].

water as in the case of CO$_2$. If the pores are filled with water, concrete carbonation becomes significantly slower. However, the presence of a certain level of moisture in concrete is the condition for the proceeding of carbonation, which is an ionic reaction. Completely dried concrete does not react with CO$_2$ and other gases at all. The dependence of concrete carbonation velocity on the relative air moisture is graphically depicted by Matoušek and Drochytka (Fig. 3).

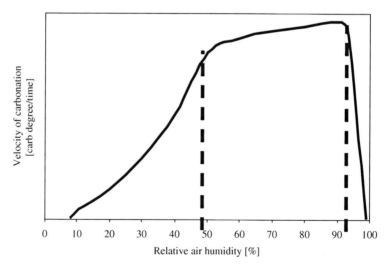

Figure 3: Dependence of the concrete carbonation on relative air humidity by Matoušek and Drochytka [1].

2.1.2 Four stages of carbonation

The process of carbonation, following long-term knowledge and measurements, has been divided into four correlated stages. The stages of concrete carbonation are very important for a correct evaluation of corrosion and the rehabilitation of the design of concrete structures.

In the *first stage* of carbonation, $Ca(OH)_2$ or its water solution in the inter-granular space is transformed into insoluble $CaCO_3$ which partially fills the pores. The main physical-mechanical parameters of concrete improve.

In the *second stage*, the rest of the cement hydration products are attacked by carbon dioxide. The products of these reactions are calcium carbonate ($CaCO_3$) and the amorphous gel of silicic acid. These substances remain in pseudo-morphologic forms after the products of binder hydration and a part of them creates very fine new crystalline formations of $CaCO_3$, respectively. New, coarse crystalline formations occur very rarely. The concrete properties do not change much; the mechanical properties fluctuate in the range of the original values.

The *third stage* of carbonation is characterised by the recrystallisation of the new carbonate formations previously formed from the inter-granular space. At the same time, many calcite and aragonite crystals that are relatively very large – ten-times and even greater – in comparison with the former systems are formed. The physical-mechanical properties of concrete gradually decrease during the third stage. A significant decrease of the concrete pH also takes place.

The *fourth stage* is characterised by almost hundred per cent carbonation. The coarse crystals of aragonite, and those of calcite, penetrate the total material structure of the cement mass. This can cause, in the worst case, a loss of the cement mass cohesiveness and strength. In this stage the pH is of course low, and a significant corrosion of the reinforcement takes place. This phenomenon can be graphically represented as seen in Fig. 4 (schema) and Fig. 5 (practical example).

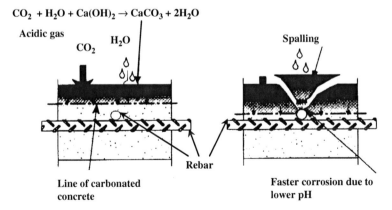

$$CO_2 + H_2O + Ca(OH)_2 \rightarrow CaCO_3 + 2H_2O$$

Figure 4: The picture represents the progress of carbonation into concrete, closer to the rebar, and consequences of the carbonation [2].

Figure 5: View of reinforced concrete wall considerably damaged by the joint effect of carbonation and sulphation. The building is in an industrial centre with high content of SO_2 in the air. (Photos without quoted names of photographer are taken from the authors' archives.)

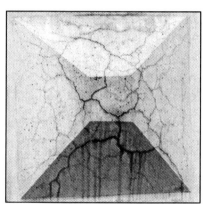

Figure 6: Typical defects of a concrete facing caused by aggregates of an alkaline reaction on the facing of a department store building with a detail of the defect (Photo: J. Habarta).

2.1.3 Alkaline reaction of aggregates

Further possible defects discovered in office buildings were in the area of alkaline dilatation (see Fig. 6). This was described in detail in Chapter 5.

2.2 Masonry structures

Masonry structure differs according to the material used for bricklaying-bricks, blocks, concrete, cellular concrete, blocks from alternative materials etc.

The surface treatment of masonry structures is the most sensitive part of these building structures. An attack by the surrounding shows itself first on facing walls.

The damage caused by water takes place mainly in the following ways:

- Water contained in the pores of the material at temperatures below the freezing point turns to ice, and forces formed during crystallisation damage the masonry structure.
- Water containing soluble substances – crystallisation during the drying of water – creates pressures that damage the plaster and the masonry, which is also referred to as efflorescence.
- Water washing out the binding components, soaking and decreasing the strength of materials, neutralisation reaction of acid rains and lime binder – all lead to the decrease of strength of the masonry (mortars, plasters, masonry from cretaceous marly limestone, limestone and sandstones with lime cement).
- Different types of volume changes – dilatation is determined by the presence of water.

2.2.1 Brickwork

As mentioned earlier, the most frequent damage of brick masonry is caused by water and by moisture in general. The brick body is relatively resistant against other corrosive environmental factors (CO_2, SO_2 solid and liquid materials such as oil products, etc.). A detailed description of the frost effect is given in Section 3.2 in Chapter 7 and the damage caused by crystallisation pressure of salts is described in Section 3.3 in Chapter 7.

The problem of mortars, and masonry mortar durability, is discussed in detail in Section 4 and especially in the Section 4.1.1 of Chapter 7.

2.2.2 Cellular concrete structures

Cellular concrete is a progressive, light building material that considerably surpasses the utilization of cellular aggregates and of concrete manufactured with these aggregates. The building systems YTONG and HEBEL are the most well-known producers of cellular concrete at the present time. They are used especially for their excellent heat-insulation properties, the favourable relationship between mechanical strength and volume mass, superior workability and more particularly for the completeness of individual building elements and the subsequent high productivity of construction. Cellular concrete has found a way into residential, agricultural and even industrial construction, especially as blocks, insulation plates or as wall components for the outer jacket of buildings. Cellular concrete represents worldwide 10% of the market of bricklaying materials, whereas in the Czech Republic it claims 25–27%.

On the other hand, there are many constructions using cellular concrete showing a typical system of failures. The failures of structures made from cellular concrete blocks are mostly caused by:

- chemical aggression of the environment,
- efflorescence,

- volume changes connected with shrinkage and a constant moisture content,
- use of unsuitable mortars and/or mortars with great thickness,
- utilisation of plasters and paints of unsuitable technical parameters, especially application of badly permeable paints and plasters by diffusion on 'fresh' masonry, and
- unsuitably designed connection of the bearing frame with the walls filled with cellular concrete.

2.2.2.1 Chemical corrosion of the environment The durability of cellular concrete, in the strict sense of the word, is chiefly understood as the frost resistance of cellular concrete and its resistance to atmospheric effects (acid rain, gases CO_2, SO_2 and H_2S). The influence of corrosive substances is increased by the macroporosity of cellular concrete, making the deep penetration of the above-mentioned substances into the material easily possible. On the other hand, this macro-porosity improves resistance against new formations of carbonation or sulphation because it enables the growth of new crystals in the pores. The velocity of atmospheric corrosion in the case of both gases is significantly influenced by the moisture of the building material (see Fig. 3 and the description of carbonation in Section 1 in Chapter 1).

Moisture, and the frequent addition nowadays of gypsum in cellular concrete production can lead to the formation of thaumasite $3CaO \cdot SiO_2 \cdot CO_2 \cdot SO_3 \cdot 15H_2O$, which by its volume pressure in the first phase, can cause the formation of micro-cracks in the masonry. The formation of thaumasite was observed mostly at the places of loading joints, i.e. in mortars with the Portland clinker.

Another chemical factor influencing cellular concrete durability is the formation of efflorescence. We know cases during the reaction with the cellular concrete mass in the autoclave when feldspar with high alkali content formed easily soluble compounds, which were subsequently leached. By the influence of the diffusion gradient these compounds gradually appeared on the surface, in spite of the properly applied surface treatment, forming crystals with dimensions up to 20 mm.

2.2.2.2 Volume changes In silicate building products we observe two types of volume changes: hydration-change and post-hydration change. In the case of cellular concrete, only post-hydration changes take place. The most frequent volume changes are connected with drying and moistening. These are chiefly reversible changes of drying from the state containing the production of moisture of 30% to the constant moisture of about 6%. These changes, following valid standards, can be the cause of shrinkage of 0.15–0.20 mm/m. Owing to the influence of these changes in the so-called 'unripe buildings', i.e. buildings with high moisture content, destruction can take place that manifests itself mainly externally on the facing. Typical examples of plaster damage can be found in Fig. 7. The failures were caused at first by the formation of thaumasite in the loading joint between blocks, and successively by moisture problems with diffusion-proof painting, which was implemented in November. The enclosed moisture after the first frosts damaged the whole structure.

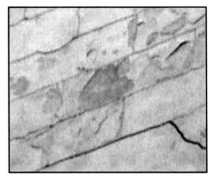

Figure 7: On the left can be seen a typical defect of cellular concrete. As the result of volume changes wall-cracking took place and cracks appeared even on the façade – typical drawing of individual wall blocks. The defect on the right is caused by volume changes resulting from wall drying. The next set of problems are defects of the wall-surface layer by frost because of the high diffusion resistance of this layer and the prevention of drying which caused the freezing of surface layers.

2.2.3 Concrete structures

Concrete structures from plain concrete are not a common material for building structures. Plain concrete is used for the manufacture of blocks with thermal insulation for outer building jackets, without thermal insulation, different slope components and grassing blocks, interlocking pavements, and profiled pavements (Figs 8 and 9). This type of material is more resistant, due to the absence of reinforcement, which is more sensitive to corrosion; carbonation for instance is not such a serious problem as in the case of reinforced concrete. The greatest problem becomes frost resistance and resistance against defrosting salts of tiles and of horizontal surfaces (Figs 10 and 11). Sufficient density of the concrete and the application of aerating admixtures provide protection against these most frequent factors.

Figure 8: Concrete block.

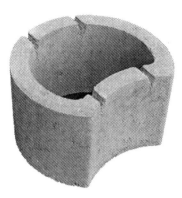

Figure 9: Slope block.

Figure 10: On the left, general view of concrete paving damaged by dilatation of the concrete base – bulging and crushing of individual tiles. On the right, a detail of concrete base-damage of the pavement.

Figure 11: Damage of the concrete pavement surface by the effect of frost and of defrosting salts causing dilatation.

Concrete blocks are to a great extent used for building of retaining walls and sloping abutment walls. The material of these structures is exposed to both rainwater and underground water. Furthermore, the influence of biological disturbances of concrete by roots of plants is not negligible.

3 Environmental damage to industrial buildings

Besides the imperfections originating in production/construction, the most frequent causes of damage and failure of industrial buildings and structures in industrial zones arise from atmospheric deterioration. In the environment around industrial agglomerations, sulphur dioxide SO_2 and other corrosive gases especially cause the deterioration of concrete.

The process of concrete deterioration by sulphur dioxide is called sulphation [3]. Sulphur dioxide (SO_2) is a markedly more corrosive gas than CO_2, but owing to its lower concentration in the normal atmosphere, the manifestation of sulphation is less frequent than that of carbonation (see Fig. 12). In general, at a lower concentration of SO_2, just as during carbonation, the effect of the relative humidity of the environment is dominant, whereas at a higher concentration of SO_2, the value of relative humidity is not as decisive. The effect of sulphur dioxide on concrete structures in mutual synergism may be seen in Fig. 12.

The process of sulphation may be clearly seen in the pictures of SEM (electron microscope) (Fig. 13).

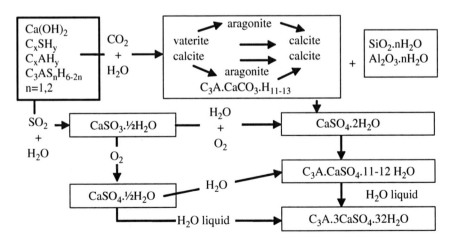

Figure 12: Simplified schema of final products of concrete carbonation and sulphation. Nomenclature: C – CaO, S – SiO_2, A – Al_2O_3, H – H_2O [1].

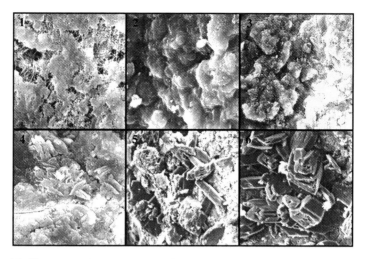

Figure 13: Figure (1) shows the original structure of hydration products in the form of fine crystals and gel structures. By the action of SO_2, first a coating is formed on the concrete surface (2), and gradually small fine crystals of new forms are created (3). Then, almost coarse crystals of gypsum $CaSO_4 \cdot 2H_2O$ (5 and 6) are gradually formed from small crystals of hemihydrates (5 and 6) (Photo: author).

3.1 Reinforced concrete framing and concrete halls

The durability of industrial reinforced concrete beams is usually highly influenced by the environment and character of production in concrete halls (see Fig. 14).

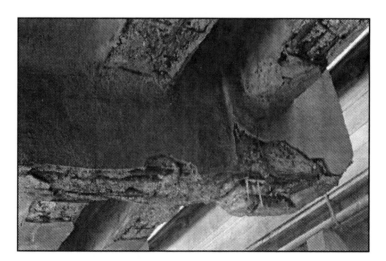

Figure 14: View of the horizontal elements of concrete skeleton monolithic hall in the chromium section of the production plant (Photo: J. Bydzovsky).

Figure 15: Details of a strongly damaged concrete layer through the action of a high concentration of sulphur dioxide and liquid moisture (Photo: J. Bydzovsky).

The most frequent type of damage of the concrete beams, of both pillars and girders, is caused by gaseous pollutants, especially by sulphur dioxide. Besides the above-mentioned effects of these gases, the pH in the given environment is also decisive in the increase of deterioration. A typical example of damage is one in which the surface of load-bearing elements of the framework monolithic hall is strongly deteriorated by sulphur dioxide, and wherein samples of the concrete surface layer minerals such as gypsum and ettringite were identified. The formation of ettringite was triggered by the access of liquid water from the process of production by the wetting itself and also by the condensation on the structure surface (Fig. 15).

3.1.1 Structures of alumina cement

From the year 1930 to the 1960s, alumina cement was abundantly used for structural concrete. Alumina cement was especially used where a rapid increase of concrete strength was required, or where winter concrete placing connected with the excessive development of hydration heat was carried out [4]. A rapid increase of initial strength is often achieved at the level of ordinary Portland cement of the same class after 28 days. However, it is generally known that through the course of time, concrete containing alumina cement gradually loses its physical-mechanical parameters to such an extent that destruction of structures very often occurs. In its composition, alumina cement markedly differs from Portland cement. Compared with calcium silicates contained in Portland cement, alumina cement most noticeably contains calcium aluminates, while the silicate components are harmful here.

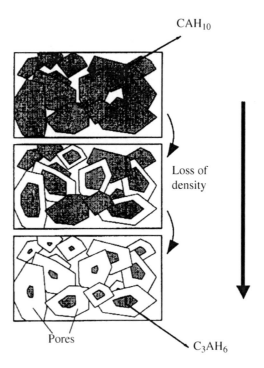

Figure 16: Conversion of alumina cement.

The decrease of alumina cement concrete strength is usually caused by its spontaneous conversion, and it is explained by a gradual conversion of hexagonal hydroaluminates CAH_{10} and C_2AH_8 formed by cement hydration to cubic hydroaluminates, C_3AH_6 which, however, take only 47.1% of the original volume (see Fig. 16). This circumstance, just as water separation, causes increase of porosity, creation of cracks and a considerable decrease of concrete strength. The increased porosity of concrete results in a more distinctive deterioration by corrosive gases and vapours from the atmosphere. Alumina cement is not used today as a material for structural purposes, but due to its specific properties it is used with an advantage as, for example, a binder for lining resistant to the action of high temperatures.

One of the greatest disasters in recent years was charted in a single aisle hall in the Czech Republic. Two of the concrete columns of 500-mm diameter were made of concrete containing alumina cement. The gradual spontaneous conversion and subsequent secondary carbonation caused an enormous decrease of compressive strength from the original 25 MPa to the value of 3 MPa (see Fig. 17).

3.1.2 Floors of concrete halls

Another potential cause of floor failures in industrial halls may arise from the effect of the alkali reaction of the aggregate and from the increased humidity. A case was discovered in which the aggregate in concrete was at adequate level, but owing to the condensation of moisture on the ceiling structure coated with paint with a high

Figure 17: View of the collapsed bearing column of the industrial hall in Uherské Hradiste, CZ (Photo: M. Matoušek).

alkali content, the alkali dissolved and dripped down to the fresh concrete floor. Then, the increased concentration of alkali caused a subsequent destruction of the concrete surface where a pool containing this solution was formed.

3.2 Cooling towers

Cooling towers represent special industrial structures from the viewpoint of service life and durability of concrete. There are two main types of cooling towers: those with a natural draught of Itterson type and towers with a forced draught (see Fig. 18). The problems of durability of both types of towers are almost identical. The structure of the cooling tower is a very thin concrete shell structure (only 100 mm in the narrowest place of the jacket thickness) exposed to extreme deterioration factors owing to its function. In lessons dealing with concrete durability, this type of structure is given as an example from the viewpoint of the comprehensiveness of action of particular factors.

In the following sections, we shall try to give a brief explanation of factors that may be encountered within cooling towers (henceforth CT), and which, especially by synergic action, cause significant failures of these towers. These factors may be classified as follows:

- the orientation of CT to cardinal points – effects of temperature and wind,
- the operational routine of CT in view of the time and especially the frequency of shutdowns in the winter period,
- the concentration of acid-forming gases in the atmosphere,
- the effect of the structure itself,

Figure 18: General view of four typically repaired draft cooling towers of the Itterson type; the structures and the interior equipment of the towers are supported by concrete columns (Photo: J. Bydzovsky).

- the poor quality of the existing concrete of the cooling tower cladding,
- insufficient cover layer of the reinforcement or the insufficient amount of the reinforcement,
- geometrical shape imperfections, and
- the absence or the low quality of secondary protection.

3.2.1 Orientation to cardinal points

Damage to the shell structure of the cooling tower is affected by the orientation to cardinal points. These factors include the effect of the prevailing wind; especially in winter months freezing and thawing of CT cladding occur. Insolation has another negative effect in which the CT cladding is time-variably heated from one side, which may cause the motion of the freezing zone (see Fig. 19). In the normal operational routine, temperatures in winter may significantly fluctuate due to different temperatures in the night and the insolated surface in the daytime. Insolation also results in the thermal expansion of the shell as a whole, during which cracks are formed that have dire consequences.

3.2.2 CT operational routine

Within the framework of operating CT in the winter season, cooling towers are shut down at very low temperatures. This causes total freezing of the wetted cladding of the cooling tower within several hours, and the structure is considerably frost-loaded because the temperature in the inner space of the tower is 16°C and the relative

Figure 19: View of a core in the cooling tower wall. The arrow shows the freezing zone, which is more damaged, compared with other parts of the cross section.

humidity is almost 100%. On the other hand, the temperature of the surrounding environment drops below the freezing point (under winter climatic conditions of Central Europe, commonly to −10 to −15°C). Therefore, at the moment of the shutdown, marked volume changes occur as a result of the creation of ice in the mass of the thin cladding of CT. This phenomenon may cause the formation of freezing zones (see Fig. 19).

3.2.3 The action of acid-forming gases from the atmosphere

The phenomena of carbonation and sulphation together with the co-action of frost and humidity effects considerably decrease the service life of the cladding of cooling towers in power plants of any type. The durability of concrete structures is markedly influenced by the action of atmospheric carbon dioxide, and in some localities also by sulphur dioxide.

The extent of concrete deterioration increases with the rise in humidity. The degree of relative humidity influences the velocity of chemical reactions between corrosive gases and products of hydration of cement, i.e. the rate of carbonation and sulphation (see Fig. 3). In a tower that was not repaired, a relatively high degree of carbonation was detected (see Fig. 20). Together with the extraction of the cement matrix, both factors contribute to the reduction of pH. Concrete with a pH value less than 9.6 loses its alkali passivation capability [5, 6].

Figure 20: View of the upper part of a cooling tower wall that deteriorated owing to the effects of carbonation, humidity and insufficient compactness.

3.2.4 The effect of structure quality

These effects are especially characterised by their dangerousness in synergic action with other effects. The poor quality of the existing concrete of the CT cladding may be caused by a number of factors such as:

- high porosity of concrete,
- high absorption of concrete (sometimes as much as 10–12%),
- low resistance to frost ('freeze-thaw resistant' concrete exhibits absorption below 6.5%),
- insufficient quality of concrete from the viewpoint of strength characteristics,
- a higher capacity to diffuse water vapour and acid-forming gases (CO_2, SO_2, etc.),
- the influence of the technology used on the cooling tower durability,
- incorrectly executed or cured construction joints,
- insufficient covering of reinforcement with concrete affecting the service life of the reinforcement,
- washing out of concrete, and
- too stringent reinforcement in the CT cladding.

In older times, the design of the construction of the majority of towers used concrete of C 16/20 type – the required strength of concrete is 20 MPa – without further requirements regarding water tightness or the durability of concrete. Neither any type of water tightness nor the number of freezing cycles or resistance of surfaces to the action of water and frost was required [6]. The concrete for use in a given environment of extremely loaded structures of cooling towers (temperature

Figure 21: View of a cooling tower wall where, owing to the marked effects of carbonation, humidity and less compactness, the concrete crumbled and the so-called 'windows' originated.

changes and also the action of external forces by the action of wind) should be designed with special consideration of structure durability (Fig. 21).

3.2.5 Geometrical shape imperfections

Geometrical imperfections of the cooling tower's cladding in the shape of local bulging, a ring or in other shapes represent a very interesting failure. Large geometrical imperfections of cooling tower cladding are especially typical in older towers. These shape imperfections are often apparent with the naked eye; however, only in a few cases they were precisely measured. In some towers, local deviations from the correct geometry reach up to almost 500 mm. Nevertheless, these large deviations, these imperfections, do not substantially affect the total load-carrying capacity and dynamic behaviour of the cooling tower. However, the danger of shape imperfections lies in the fact that in the place of imperfections, peak local bending moments originate for which the cladding was not designed. In these places, cracks in concrete occur before the tower is put into operation. These cracks, moreover, extremely accelerate the process of permanent deterioration of the concrete's condition. Shape imperfections are also connected with the fluctuating and generally insufficient cover layer.

3.2.6 Protective surface finish of concrete

In the majority of cooling towers, a protective surface finish was not applied during their construction. This fact is still apparent only on the cooling towers that have

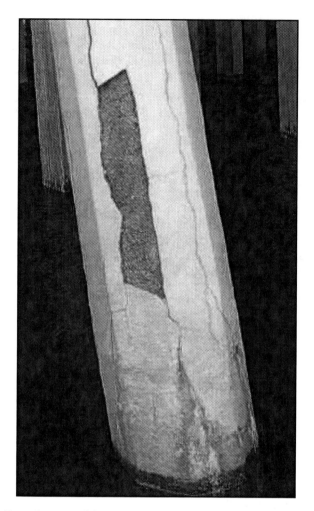

Figure 22: Frost damage of the concrete cover of stanchions (columns) supporting
the whole structure of a cooling tower. The stanchions are fixed-ended
in a foundation plate flooded with water.

not been repaired. Surface protective systems form an additional barrier against
the penetration of undesirable media on the surface of the rehabilitated concrete
structure and in this way prolong its durability, and especially, help the passivation
of the steel reinforcement (Fig. 22).

The secondary surface protection of the concrete structure is not carried out using
only one coat of paint, but a comprehensive surface protective barrier composed
of several layers where each of these layers may secure different properties (e.g.
penetration guarantees water tightness, the cover layer the diffusion capabilities,
etc.). As far as the corrosive media are concerned, these especially refer to the
penetration of CO_2, SO_2 and H_2O, either as vapours or as liquids.

Figure 23: Completely damaged and non-functional paint of the interior part of the tower (Photo: R. Kepák).

In cooling towers, the interior surface finish (see Fig. 23), which should prevent the penetration of water vapour and water from the tower's interior environment (100% humidity) into the structure, is of greatest importance. And conversely, the exterior surface finish should allow water vapour to escape from the structure (fully vapour-permeable paint), but at the same time, not allow water (rainwater) into the structure. Then, the whole surface finish should simultaneously limit the access of carbon dioxide or other corrosive media into the concrete structure from both sides.

3.3 Wastewater treatment plant, sewerage system and pits

In the field of water management structures we may come across water treatment plants that usually either comprise water reservoirs, drinking water tanks, etc. or have facilities for wastewater. The main feature of water treatment facilities is that they are mostly designed for the addition of various chemicals that are of low concentrations, so that their harmful activity is not too distinctive. Structures for this purpose utilise impermeable concrete, which is a guarantee for the trouble-free operation of the whole treatment plant. Only certain cases were detected when during the repair of the treatment plant the silicate paints based on water glass reacted due to chemical additives in water, especially with the addition of chlorine.

Far more distinctive corrosion activity may be seen in wastewater. In general, wastewater is divided into the following four main groups:

1. municipal wastewater,
2. industrial wastewater,
3. rain water (storm water), and
4. infiltration subsurface water.

3.3.1 Municipal water

Urine and faeces make the principal part of polluted municipal water, almost 80% of organic substances. Faeces dry matter is formed by the residua of intestinal bacteria, lipids, proteins, polysaccharides and products formed in their decomposition. The largest proportion of organic substances form nitrogenous substances is mostly represented by urea amino acids and ammonium nitrogen. However, in municipal water, biological hydrolysis of urea and biological decomposition of amino acids occur very quickly. In inorganic substances, sodium and chlorides are dominant. The concentration of chlorides is especially very high. Then follow the sulphates (forming as much as 80% of the entire sulphur content) and phosphates (phosphorus – predominantly present in the inorganic bonded form, of which 60% is in soluble form) (see Fig. 24).

The temperature of the municipal water in the sewerage depends on the season. In winter the temperature ranges from 8 to 12°C. In summer it approaches 20°C. Biochemical processes are accelerated with the temperature increase, so the dissolved oxygen is depleted more easily. The reaction of municipal wastewater (see Fig. 25) is normally slightly alkaline; pH is usually within the range 6.5–8.5 [7].

Figure 24: View of completely damaged spillway wall of the pit through which the sludge from municipal water overflows (Photo: J. Hromádko).

Figure 25: View of a completely damaged wall of one of the pits in a wastewater treatment plant (Photo: J. Hromádko).

3.3.2 Industrial wastewater

This kind of wastewater refers to water discharged into the sewerage from industrial plants, agricultural wastewater or water pre-treated in the plant. Industrial wastewater, unlike municipal water, is of varied character. In inorganically polluted wastewater, the substance present may either be in un-dissolved or dissolved form, and may rank among toxic or non-toxic substances. Industrial wastewater may be divided into the groups:

- wastewater predominantly polluted inorganically by un-dissolved substances (wastewater from coal washing, ceramic and glass industry,
- wastewater predominantly polluted inorganically by non-toxic dissolved substances (see Fig. 26) (wastewater from plants producing potash, phosphate and nitrogenous fertilisers, an iron preparation plant, soda production, agriculture industry), and
- wastewater predominantly polluted inorganically by toxic dissolved substances (surface metal treatment, radioactive water, etc.).

Organic substances contained in industrial wastewater may also be either un-dissolved or dissolved when divided into four groups:

Figure 26: Concrete deterioration of the catch basin in a chemical plant.

- non-toxic and biologically decomposable substances,
- non-toxic and with difficulty biologically degradable substances,
- toxic and biologically decomposable substances, and
- toxic and with difficulty biologically decomposable substances.

3.3.3 Rainwater
This kind of wastewater is discharged from the municipality area through the public sewerage. The quality of precipitation water (rainwater, melting snow) is very unsteady and depends on many circumstances. The concentration of organic pollution in this water is like that in municipal wastewater. In winter, the precipitation water from melting snow absorbs a large amount of salts from road maintenance, which demonstrates itself in a large increase of concentration of chlorides. Rainwater can transport large amounts of inorganic material which can cause mechanical abrasive wear (see Fig. 27).

3.3.4 Infiltration underground water
Infiltration underground water, the so-called ballast water, is infiltrating into the sewerage system from the surrounding soil. This kind of water is usually not much polluted by corrosive ions, which depends on the surrounding soil. It is a bigger problem for wastewater treatment plants due to the thinning of wastewater.

Figure 27: Extreme case of the effect of extracting rainwater. Concrete wall damaged by water flowing out for a long time from a broken pipe discharging precipitation water; poor compaction also had an effect.

4 Environmental damage to transportation infrastructure

Transportation infrastructure, i.e. bridges, roads, retaining walls, tunnels, etc. are exposed to the action of a number of specific effects (e.g. de-icing substances), which may cause significant failures.

4.1 Bridges and pavements

Besides other corrosive phenomena affecting all reinforced concrete structures, the deterioration caused by chlorides (see Section 2.3.3 in Chapter 5), alkali expansion (see Section 3.1 in Chapter 5) and stray current are especially dangerous for bridge structures, particularly due to the use of pre-stressed reinforcement.

4.1.1 Stray current
The Pourbaix diagram of conditions of electrochemical corrosion shows the dependence of the potential on pH for the iron-water system. However, these processes and conditions substantially differ with the presence of foreign ions and an exterior electric field. The corrosion rate may increase many times if electric current is introduced. This electric current may be represented by stray earth current that penetrates into the structure from a foreign source. Stray earth current may pass through railway sleepers and structures in the vicinity of places where direct current is used. Structures that face the greatest threat represented by stray earth current are in the vicinity of electrified railway lines using direct-current traction voltage 1.5 and 3.5 kV. So as to limit stray earth current, the rails are welded or provided with

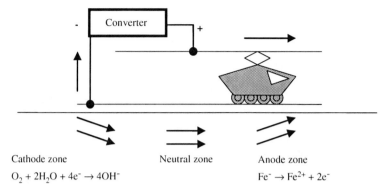

Cathode zone   Neutral zone   Anode zone

$O_2 + 2H_2O + 4e^- \rightarrow 4OH^-$ $Fe^- \rightarrow Fe^{2+} + 2e^-$

Figure 28: One way of possible protection of reinforced concrete structures –
correct wiring of stray current circuit.

welded current-carrying inter-connections that ensure their proper and permanent
current connection (see Fig. 28).

In the case of structural failure where one of the following conditions is satisfied,
an exploration must be carried out to discover stray earth current occurrence:

- the direct-current traction railway line is at a distance of 5 km;
- up to a distance of 500 m from the structure there are structures which are
  potential sources of stray current – the transformation station, facilities of gas
  pipelines and water piping and cables;
- geological subsoil with the potential of becoming sources of spontaneous polarisation, deposits of ores, graphite or tectonic zones.

The exploration of the occurrence of the stray current should include the
following:

- the determination of the direction and intensity of the electric field, the measurement of the reinforcement-soil potential,
- the measurement of the earth resistance of supports and the load-carrying
  structure,
- the measurement of the potential gradient, and
- the measurement of the insulation resistance – if enabled by the structural design
  and the scope of the repair.

The situation of existing structures is worse owing to the contamination by conductive ions; the reinforcement is usually not bonded, and the electric resistance
between the load-carrying structure and the substructure is low.

The increase of the electric resistance of the reinforced concrete structure to earth
ranks among the basic protective measures. In existing structures, only secondary
protection by coatings, paints, sprays, foils, insulation strips and the structural
design (the separation of particular parts electrically) is possible.

Figure 29: View of a road where concrete failure by alkali expansion was detected.

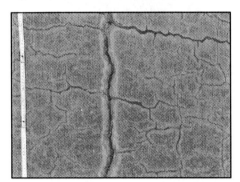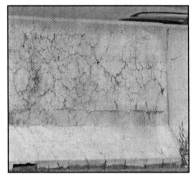

Figure 30: To the left – a detail of the airport concrete runway, and to the right –
the crash barrier, indicating the same cause of failure.

4.1.2 Volume changes – alkali reaction of aggregate and frost failure

Failure due to the alkali expansion of the aggregate is another serious failure of concrete structures (see Section 3.1 in Chapter 5). This kind of structural failure has been known since the 1930s, and it was first observed and precisely described for highways in California, USA. Based upon the existing knowledge, it may, however, be stated that the phenomenon of the alkali expansion was somewhat neglected in the past. The occurrence of this type of failure highlights the fact that the threat of alkali expansion cannot be minimised (see Figs 29–31).

4.1.3 Carbonation, effects of chlorides and humidity

The combined effect of carbonation resulting in the decrease of concrete pH and the subsequent corrosion of the reinforcement represents a significant source of damage. Extremely high corrosion activity in co-action with chemical de-icing salts – chlorides – occurs especially in bridges within a short period of time. The corrosion products of the reinforcement are of larger volume than the original reinforcement, which induces tensile stress in the concrete cover layer. This tensile

Figure 31: For comparison, an example of concrete frost damage of a crash barrier with the co-action of de-icing salts.

Figure 32: View from below of one of the longitudinal girders of a bridge. An example of the extensive reinforcement corrosion caused by the synergic action of moisture and chlorides (Photo: Jiří Bydžovský).

stress is larger than what the concrete is able to stand in tension, and consequently, cracking and even the total spalling of the cover occur (see Figs 32–34).

4.2 Tunnels

From the viewpoint of durability, concrete structures of tunnels, like other concrete structures, are endangered. The following factors especially act here to a large extent:

- liquids that react with the cement matrix in the following ways:
  - extraction of cement matrix,
  - creation of easily soluble compounds and their washing,
  - creation of compounds of larger volume.

Figure 33: Detailed view from below of the bridge arch where corrosion is caused by the synergic action of: water leaking, carbonation and chloride attack (Photo: Jiří Bydžovský).

Figure 34: View of the bridge pillar corner. The damage was caused by precipitation and re-crystallisation of the calcium constituents of concrete. The high content of chlorides in water leakage into the concrete also contributed to the extract.

- oils and petroleum products,
- increased content of corrosive gases in the atmosphere, and
- high temperatures.

With respect to the closed space and the action of the chimney effect in case of fire, the risk of damage to concrete by high temperatures is higher in tunnels, especially in reference to road tunnels (Fig. 35).

Figure 35: During a fire in a tunnel, the tunnel is completely disintegrated owing to high temperatures that may reach a maximum of 1400°C. The picture shows a warped reinforced vault made of shotcrete (Photo: J. Horvath).

During a fire, the temperature may rapidly increase within a short period of time. In such cases, concrete is damaged by the explosive spalling due to the increase of water vapour pressure in the surface layers of concrete as illustrated in Fig. 36. The situation is accentuated in high-performance concrete used in most cases for the construction of tunnels.

The pictures (Figs 37 and 38) show the damage to concrete in tunnels. Following a series of fires in road tunnels in the Alps, the problem of increasing concrete durability at high temperatures is currently being solved in Austria [8]. Researchers are endeavouring to eliminate the destruction of concrete caused by the action of high temperatures, particularly to prevent the spalling of the cover layer of concrete above the reinforcement. The warped parts of the concrete structure make access more difficult and increase problems during the rescue work and removal of the fire's ill effects.

5 Farm buildings deterioration

The deterioration of farm buildings may be major, especially in facilities where manure is handled and where fertilisers or other chemical agents are stored. It also occurs in buildings serving as animal housing due to the effect of indoor pollution and the interior climate. Another potential damage to structures may be the deterioration of their properties by biological corrosion (see also Section 2.2 in Chapter 5).

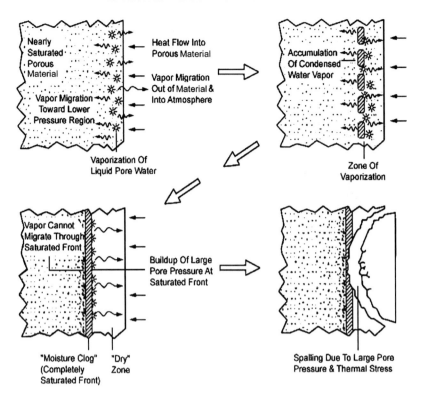

Figure 36: Diagram of the origin of damage to concrete surface layers during fire [8].

Figure 37: General view of a tunnel vault without concrete surface layers damaged by fire (Photo: J. Horvath).

Figure 38: Detailed view of the tunnel vault without concrete surface layers damaged by fire (Photo: J. Horvath).

5.1 Corrosion effects during storage of fertilisers

Nitrogenous, phosphate and magnesium-phosphate fertilisers rank among the most commonly used fertilisers. These are generally powdery or crystalline substances that, when applied or handled, strongly raise dust and stick to the concrete surface. This particularly refers to the hygroscopic substances that with atmospheric moisture create strongly corrosive substances such as $NaCl$, NH_4Cl, and $(NH_4)_2SO_4$ (see Section 1 in Chapter 7). The consequences of corrosion caused in this way are observable as early as after 2–3 years. It is a case of different type of deterioration, especially the extraction of the binding matrix as a result of the chemical reaction, or cover layer spalling and uncovering of the reinforcement, which afterwards is quickly subject to further corrosion (Fig. 39).

5.2 Micro-climate and indoor pollution of stables and corrosive solutions

The micro-climate of animal housing structures and also of some storehouses is an important corrosion factor. The construction of large-scale installations, the concentration of animals in this space and difficulty in ventilation create an adverse micro-climate for the structures. This especially refers to higher humidity, and the occurrence of sulphate and ammonia. Owing to high relative humidity, its condensation and the dilution of gaseous products that originate from the biological processes of excrements of stabled animals and also from their breathing, diluted solutions of acids are created. These are particularly H_2CO_3, HNO_3 and H_2SO_4. These acids act corrosively on all surfaces including ceilings and walls. The corrosive acid H_2SO_4 also originates by the oxidation from H_2S, formed by the bacterial decomposition of organic substances.

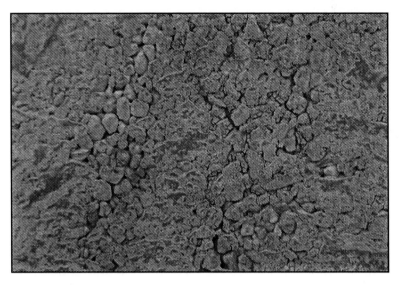

Figure 39: Detail of the interior part of the damage to a granulation tower – cement matrix extraction.

Besides animal housing structures, problems with a corrosive environment may also occur in other places, e.g. in storehouses for potatoes where the temperature is maintained at 2–4°C and the humidity at about 90%. In such facilities a high concentration of SO_2 or H_2S was found. SO_2, on the one hand, causes concrete sulphation and, on the other hand, hydrates to corrosive aqueous solutions of $SO_2 - SO_2$ and $6H_2O$ [9].

Corrosive solutions from different farm products or wastes represent other corrosive media. One of these are, e.g. silage juices – turbid dark-coloured liquids of pH 3.5 that originate from anaerobic lactescent fermenting processes. These substances contain a relatively large amount of saccharides and acids (lactic acid, butyric acid, acetic acid). Silage pits are constructed with non-reinforced concrete, which better resists corrosion, but nevertheless, the damage used to be frequent even after several years as a result of frost effects, variations in temperatures and humidity. So that the structure may be in the long term usable, it must be constructed with dense well-processed concrete or provided with painted surfaces.

Mechanical damage to concrete by biological effects is not frequent, but trees or other plants may damage concrete by their roots in the cracks of the material. The growth of plants such as moss (see Fig. 40), lichens and algae on concrete surfaces causes little mechanical damage to the concrete surface itself. Rather, the harmful effect is represented particularly by the fact that these plants retain water on the surface. Then, the cause of the damage is the crystallisation pressure at subzero temperatures.

Bio-corrosion through the action of bacteria consists in the chemical action of the products of bio-chemical processes, e.g. products of sulphur. Bacterial damage

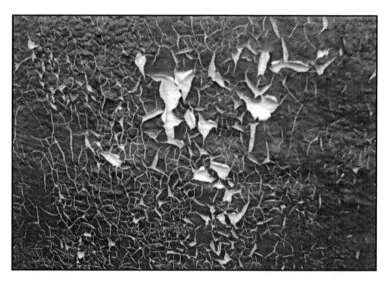

Figure 40: This picture shows an interesting form of damage to the paint of the interior part of stables where the paint is covered with a thin layer of organic residue. For operational reasons, there is a cyclic desiccation of the environment, which results in the development of the tensile stress that causes the separation of the paint from the base. The adhesion of a biological material to the paint surface is larger than that between the paint and concrete.

to concrete used to be time-limited, which leads to the situation where the potential origin of the damage by bio-corrosion is not taken into consideration at all.

5.3 Bio-corrosion of concrete

• Sulphate attack: Sulphur bacteria are contained in stagnant water, soil and in all places where sufficient sulphur and water is present. Sulphur bacteria in the presence of O_2 in water oxidise hydrogen sulphide (H_2S) to corrosive H_2SO_4. On the contrary, under the condition of the absence of air access, aerobic micro-organisms take oxygen from sulphates dissolved in water and reduce dangerous H_2SO_4 to a relatively harmless H_2S. These bacteria are especially active in the sewerage system. After 5 years of exposure, concrete can be attacked up to 30 mm of depth (2–3 pH) [10].
• Nitrification: A similar action based on nitrogen decomposition.

6 Deterioration of other structures

This section describes the deterioration caused by the environment of structures other than those mentioned in the previous sections.

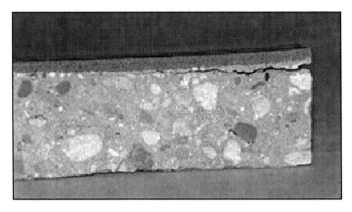

Figure 41: Detailed view of a section of a damaged epoxide floor. The separation of the polymer floor from the concrete base resulted from the trapping of moisture in the base and the cyclic change of temperature. The overall spalling was due to the de-icing of the screeding under the paint, and a role was also played by the deterioration of the polymer material by UV radiation (Photo: M. Vyhnánek).

6.1 Damage to materials containing polymers

Materials utilised in the construction industry which contain polymers may be divided into three groups [11]:

- purely polymer materials (products of plastic materials, polymer surfaces – floors),
- materials in which a polymer is used as a binder (polymer concrete – PC), and
- polymer is used as an additive (polymer-cement concrete – PCC).

Owing to the relatively higher resistance of polymers to different corrosive environments and chemical agents, the damage is usually not too distinctive compared with silicate materials. The diversity of polymer materials used in the construction industry is so large that in general, it is not possible to specify their corrosion resistance. Generally, polymer materials do not resist higher temperatures and are prone to ageing due to UV radiation. The majority of failures in structures used to be caused by improper execution or application under inappropriate conditions, e.g. by the sensitivity of most polymers to humidity or insufficient temperature for polymerisation.

As far as the negative action of water and temperature on coating materials is concerned, the damage caused by polymers may be classified according to the following mechanisms:

- When water is trapped under the layer of a building material together with a polymer, water gradually changes into vapour that afterwards induces tensile

Figure 42: Detailed view of the gallery of the top part of a cooling tower exhibiting damaged red and white painted stripes based on epoxy resins. When this kind of paint is used, overall spalling occurs due to the damage of the concrete under the paint caused by the freezing-thawing action. A role is also played by the deterioration of the polymer material by UV radiation.

stress (e.g. due to the insolation of a wet structure provided with a non-permeable paint) (see Fig. 41).

- When water is trapped under the layer of a building material together with a polymer, and water condensation occurs, and owing to low temperatures water is changed into ice.
- Mixing a wet aggregate with a binding polymer component during the production of polymer concrete causes the slowdown or even the prevention of polymerisation (also, the moisture condensation on the aggregate's cool surface is prevented).

6.1.1 Paints

Besides the effects on the durability of polymer materials mentioned above in paint systems, the following inter-connected effects may be designated [12]:

- insufficient thickness of the paint system makes water penetration possible, but conversely, this prevents its evaporation,

- lower water tightness and together with frost induced damage of the whole system independently or even with the base, and
- high diffusion resistance prevents evaporation, increases water condensation under the paint and prevents subsequent secondary failures (see Fig. 42).

6.2 Piles

Besides the damage by stray current to the reinforcement in reinforced concrete (see Section 1 in Chapter 2), plain concrete may also be damaged by the transport of corrosive ions through the soil. Figure 43 shows a specific example of a pile strongly damaged by sulphate expansion (see Chapter 4, Section 2.1.1). As a result of the action of stray current, SO_4^{2-} ions from groundwater were transported through the soil as far as to the unprotected part of the structure. A part of the pile embedded in the highly corrosive groundwater was protected by a hydro-insulation foil.

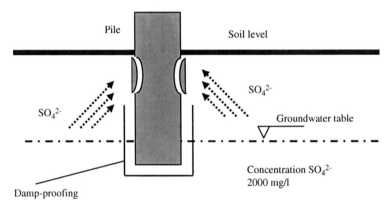

Figure 43: Chart of pile corrosion by corrosive ions from ground.

References

[1] Matoušek, M. & Drochytka, R., *Atmosfericka koroze betonu* [*Atmospherical Concrete Deterioration*], IKAS: Prague, 1998.
[2] Emmons, P.H., Drochytka, R. & Jeřábek Z., *Sanace a udrzba betonu* [*Concrete Repair and Maintenance*], CERM: Brno, 1999.
[3] Cowan, H.J. & Smith, P.R., *The Science and Technology of Building Materials*, 1st edn, Macmillan: Canada, 1988.
[4] Neville, A., *High Alumina Cement Concrete*, 2nd, The Construction Press: Lancaster, UK, 1980.
[5] Emmons, P.H., *Concrete Repair and Maintenance Illustrated*, R. S. Means Company: Kingston, MA, 1996.
[6] Drochytka, R. & Hela, R., Natural draught, cooling towers. *4th Symposium on Cooling Towers*, Kaiserslautern, 1996.

[7]   Novak, J., *Sewer Operators Handbook*, Medim: Prague, 2003.
[8]   Schneider, U. & Horvath, J., Behaviour of ordinary concrete at high temperature. *Proceedings of the International Conference on Construction and Architecture, Part 2*, National Technical University of Minsk, Minsk, Weißrussland, TU Wien, pp. 3–113, 2003.
[9]   Greenwood, N. & Earnsha, G., *Chemie prvků [Chemistry of Elements]*, Informatorium: Praha, 1993.
[10] Wasserbauer, R., *Biologické znehodnocení staveb [Biological Damage of Structures]*, ABF ARCH: Praha, Prague, 2000.
[11] Drochytka, R., Dohnálek, J. Bydžovský, J. & Pumpr, V., *Technické podmínky pro sanace betonových konstrukcí TP SSBK II [Technical Conditions for Repair Work of Concrete Structures]*, SSBK: Brno, 2003.
[12] Vyhnánek, M., Lena Chemical Company Information, http://www.lenachemical.com/gb/index.phtml, 2002.

CHAPTER 9

Environmental deterioration of timber

S. Mindess
Department of Civil Engineering, University of British Columbia, Canada.

1 Introduction

Timber structures can be remarkably durable if they are properly designed, constructed and maintained. However, wood is a naturally occurring biological material. Thus, timber structures are also subject to decay due to a number of biological factors such as bacteria, fungi, insects and molluscs, and due to such non-biological factors as weathering, wetting and drying, chemical exposure and atmospheric contaminants. Timber is as well susceptible to fire. Here, the environmental deterioration of timber structures is discussed, as well as the available preventative measures that may be taken to preserve timber. Finally, some of the economic considerations involved in timber preservation are considered.

Wood is the oldest and one of the most widely used construction materials. It combines a high strength-to-weight ratio, appealing aesthetic properties, ease of construction, the ability to be repaired, and cost effectiveness. It is also a renewable resource, and its production is much less energy intensive than that of other construction materials, such as steel, aluminium or concrete. Currently, the annual production of wood is about 10^9 tonnes, about equal to the world production of iron and steel, and about one-tenth of the world production of concrete (about 10 billion tonnes).

The terms *wood* and *timber* are often used interchangeably, but we will distinguish between them here. *Wood* will be used to refer to the basic material that we obtain from trees. *Timber* (or *lumber*) will refer to the sawn structural members (beams, planks or boards) which are used in construction. Timber will, in general, contain a large number of macroscopic defects, such as knots and cracks; the properties of timber are governed both by these macroscopic features and the underlying structure of the wood itself.

Figure 1: Liao dynasty temple at Dule, China (984 AD).

Fig. 1 shows one of the oldest timber structures in China: a Liao dynasty temple constructed in 984 A.D. The oldest existing wood-frame structure in the United States (Fig. 2) is a home built in 1636. And, on a totally different scale, some of the violins (Fig. 3) produced in Cremona, Italy, as far back as the 17th century are still in use today. On the other hand, there are all too many cases in which wood-frame structures have had to be rebuilt after only three to five years in service.

2 Wood as a material

The *macrostructure* of wood is shown schematically in Fig. 4. Briefly, the *outer bark* is a dense layer that protects the interior of the tree. The *inner bark* transports sap from the leaves to the growing parts of the tree. The *cambium* is a layer of tissue, one-cell thick, between the bark and the wood. Its repeated subdivision forms both new wood and new bark. *Sapwood* is the wood near the outside of the tree trunk; it conducts moisture up from the roots and stores food (carbohydrates) needed for further growth. *Heartwood* is the inner core of the trunk and is composed of non-living cells; it is drier and harder than sapwood, and is more resistant to decay. The *annual rings* are formed as the cells of cambium grow and divide during a growing season. During the period of rapid growth in the spring, the cells are relatively large with thin walls, and are referred to as *springwood*; the *summerwood* forms later in the growing season, with smaller cavities and thicker walls than the springwood. Finally, the *pith* is a small cylinder of primary tissue at the centre of the tree around which the annual rings form.

Figure 2: Fairbanks House, 1636-40, Dedham, Massachusetts (Photograph courtesy of Dr. Jeffery Howe, Boston College, Boston, MA).

Figure 3: Amati violin, 1677.

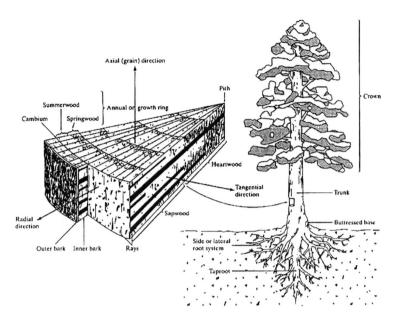

Figure 4: Schematic illustration of the parts of a tree and of the structure of wood.

On a *microstructural* level, wood may crudely be modelled as consisting of bundles of aligned, thin-walled tubes (fibres) glued together. Typically, about 90% of the volume of the wood consists of longitudinally oriented cells (*tracheids*), as shown in Fig. 5 for two different species, birch and Douglas fir. These tracheids are responsible for the mechanical support of the tree, and for vertical conduction of water and sap. The remaining 10% consists mostly of transversely oriented cells (Fig. 6), whose function it is to store the food for the tree and transport it horizontally.

Wood is an organic material. On a *molecular* level, wood consists mainly of cellulose, hemicellulose, lignin and extractives (a variety of chemical substances responsible for properties such as colour, odour, taste, resistance to decay, flammability and hygroscopicity); there are also very minor quantities of inorganic materials. The *elemental* composition of dry wood is approximately 50% carbon, 44% oxygen, 6% hydrogen and 0.1% nitrogen. Its general chemical composition may be approximated as $C_6H_9O_4$.

Finally, since wood can exchange moisture with the atmosphere quite easily, it will eventually reach moisture equilibrium with its surroundings. Moisture can exist in wood as *free water* within the hollow cell cavities, and as *bound water* physically adsorbed in the cell walls. The point at which the free water has completely evaporated while the cell walls are still fully saturated is known as the *fibre saturation point*, which occurs at a water content of approximately 27%. If wood is dried below this point, it will begin to shrink, which may lead to cracking and warping as described later. Kiln drying typically reduces the moisture content of timber to about 16%. In equilibrium with exterior air, wood generally has a moisture

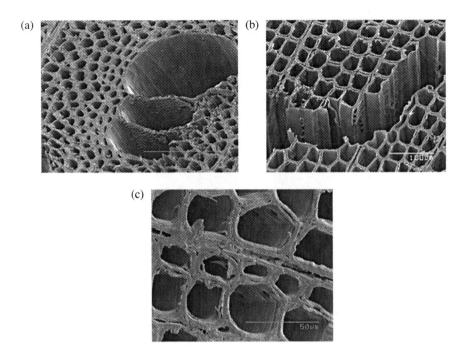

Figure 5: (a) Microstructure of birch perpendicular to grain. (b) Microstructure of Douglas fir perpendicular to grain. (c) Close-up of Fig. 5b. (Photographs courtesy of Ms. Mary Mager, Department of Metals and Materials Engineering, University of British Columbia.)

content of about 12% to 18%. In a heated building, this will be reduced to about 4% to 10%. The effect of moisture content on deterioration will also be discussed later.

Like other naturally occurring biological materials, wood is subject to deterioration over time. The factors that cause deterioration may be divided into two groups: *biological factors* and *weathering*. (It should be noted that this review deals only with the deterioration of timber in construction; biological attack on living trees will not be considered here).

3 Biological factors which cause deterioration

Though different species of wood have different degrees of resistance to biological attack, all will deteriorate under the appropriate conditions. Wood may be attacked by a number of different organisms. To thrive, these organisms require food, oxygen, moisture and warmth. The food, of course, comes from the wood itself. There is generally enough oxygen present in wood to promote the growth of these wood-eating organisms. Even fully submerged wood can be attacked by marine organisms.

Figure 6: Douglas fir, tangential plane. (Photograph courtesy of Ms. Mary Mager, Department of Metals and Materials Engineering, University of British Columbia.)

(The British warship, the Mary Rose, which was raised from the seabed in 1982 after having been sunk in a naval engagement in 1545, must be sprayed continuously with cold, fresh water to prevent further biological attack on her timbers). While keeping wood dry will prevent attack from fungi, some insects will attack dry wood. As well, most wood-eating organisms do well at the moderate temperatures which generally promote plant growth. Thus, preventing wood deterioration is generally not a simple procedure, and clearly different types of preservation treatment are required for different types of exposure. However, in general, if wood can be kept very dry and is not exposed to wood-eating insects, it may survive for a very long time, as wooden artefacts recovered from the tomb in Egypt of Tutankhamen, who died in 1352 B.C., attest (Fig. 7).

3.1 Fungi

Fungi are a class of non-flowering plants that contain no chlorophyll. They cannot manufacture their own food, and therefore can grow only on dead or living organic material. They reproduce through the production of microscopic spores that may be spread in various ways. These spores develop into thread-like *hyphae* which spread through the wood. They secrete an enzyme which de-polymerises the long-chain

Figure 7: Ebony and ivory game board of Tutankhamen.

cellulose molecules and the lignin structure, eventually softening and weakening the wood. Fungi grow best at temperatures in the range of about 20°C to 35°C. At low temperatures, fungi remain inactive, but will not die. They also become inactive at temperatures above about 45°C, but will not die until the temperature reaches at least 60°C. There are a number of different types of fungi that may attack wood, and that may cause different types of deterioration. They can often be detected visually, as they sometimes produce fruiting bodies, such as toadstools and even some types of mushrooms.

3.1.1 Brown rot fungi

Brown rot fungi cause the most damage to wood, particularly softwoods, in temperate climates. They attack primarily the cellulose (though they also modify the lignins), and the wood takes on a brown colour. Cracking across the grain, abnormal surface shrinkage and eventual collapse accompany this attack, and there is a rapid loss of strength in the affected wood. For instance, it has been reported that Douglas fir heartwood, which is fairly durable, can lose as much as 25% of its compressive strength perpendicular to the grain, and 15% of its compressive strength parallel to the grain, after only one week of exposure in conditions ideal for fungal growth. Even greater losses would occur in the less durable sapwood.

3.1.2 White rot fungi

White rot fungi, which attack primarily hardwoods, remove both cellulose and lignin from the wood. They cause the wood to lose colour, thus making it appear paler or 'whiter'. Unlike wood attacked by brown rot fungi, white rot fungi do not normally cause much cracking or shrinkage; strength loss occurs at a relatively slow rate.

3.1.3 Soft rot fungi

Soft rot fungi attack the lignocellulose in wood. They attack wood that is permanently saturated with water, such as in cooling towers, or in permanent contact with moist soil. They may also attack wood that undergoes cyclic wetting and drying over a long period of time. Soft rot generally attacks only a thin surface layer of the wood. Hardwoods are more susceptible than softwood to soft rot. This type of rot is generally less serious than either brown or white rot.

3.1.4 'Dry rot' fungi

The term 'dry rot' is a misnomer, since all wood must be damp in order for rotting to begin (Fig. 8). These fungi have water-conducting strands that can carry water (usually from the soil) to wood that would normally be dry, so that it becomes moist and subject to rotting.

Figure 8: Fungus bloom and dry rot. (Photograph courtesy of Steve Lay, Trinity Termite and Pest Control, Redding CA.)

3.1.5 Mould and stain fungi

Mould and stain fungi live on the carbohydrates stored mostly in the sapwood, and thus have little or no effect on the strength properties of the wood. They thus cause mostly aesthetic damage to the wood. (It should be noted, however, that some moulds may pose a health hazard in some cases to humans, particularly if there is a high level of mould present in a home). The stains produced by these fungi can take on a variety of colours; black, blue and grey are most common, but brighter colours such as various shade of yellow, red, green and orange also occur. On softwoods, the resulting stains can often be brushed off or surfaced off. On hardwoods, however, the stains generally penetrate too deeply to be removed easily.

3.1.6 Control of fungi

For most wood-destroying fungi, the moisture content must be above the fibre saturation point (i.e. above 25%–30%). Since kiln-dried or even properly air-dried woods have moisture contents less than about 20%, they should not be subject to rot *if they remain dry in service*. However, due to improper construction techniques, the amount of water held in homes or commercial buildings may be excessive, and condensation may occur. This can readily and rapidly lead to severe rotting damage, as in the 'leaky condominium' problems that have occurred in British Columbia and elsewhere (Figs 9 and 10). Thus, control of the moisture content of the wood is the best and easiest method of controlling fungi. This may involve proper ventilation, ensuring that no plumbing or other leaks occur, and sealing the wood with paint or other waterproofing materials. Chemical treatments of the wood for this purpose will be discussed later.

Figure 9: Damage to leaky condominium in Vancouver, Canada, due to rotting wood. (Photograph courtesy of Levelton Engineering Ltd., Richmond, BC.)

Figure 10: Rot observed in wall during recladding of structure. (Photograph courtesy of Levelton Engineering Ltd., Richmond, BC.)

3.2 Bacteria

Bacterial growth may occur in wood that has a high moisture content – freshly cut, in contact with damp soil, or stored in water or under a water spray. While bacterial decay is not generally a major problem, it may cause a significant loss in strength over long periods of times (decades). Some bacteria may increase the permeability, or the absorptivity, of wood.

3.3 Insects

There are a number of different wood-destroying insects that may infest wood, including termites and some species of beetles and ants. Collectively, they are responsible for a great deal of damage.

3.3.1 Termites

Termites are social insects that live in large colonies (Fig. 11). There are three main classes of termites: subterranean termites, dampwood termites and drywood termites.

Subterranean termites are responsible for most of the serious wood damage. They live in underground nests in damp soil, and will infest wood either directly in contact with the nest, or which they can reach through mud tubes (up to 100-m long) that they construct. The 'worker' termites excavate tunnels and chambers through the wood; it is the formation of these galleries that weakens the wood.

Figure 11: Termite.

One-celled protozoa in the digestive tracts of the termites then break down the wood cellulose from the wood so removed into suitable 'food'.

Dampwood termites generally nest above ground in damp or decaying wood. They require a high moisture content in their nesting area, but can invade adjacent dry wood as well. Compared to subterranean termites, they are of minor economic significance.

Drywood termites can live in dry wood, without contact either with the soil or other sources of moisture. They multiply less rapidly than subterranean termites, and do not destroy wood as rapidly. However, over time they too can do a great deal of damage.

3.3.2 Control of termites

Apart from chemical treatments, which are discussed later, the best protection against termites is to prevent them from gaining access to the structure. This may involve ensuring that the timber parts of a structure are not in contact with damp soil, preventing cracks in concrete slabs, providing proper ventilation, and so on. Proper painting will generally prevent the ingress of drywood termites.

3.3.3 Carpenter ants

Unlike termites, carpenter ants do not use wood as a food source. Rather, they burrow into wood in order to create nesting sites. They may also nest in insulation adjacent to the wood. If left unchecked, they can cause serious damage in structural members as the colonies grow or move to different parts of the structure. As with other insects, carpenter ants prefer to locate close to a source of food. Thus, the best means of prevention is to maintain a sanitary environment, avoid moisture problems around or within the structure, and prevent vegetation from coming into contact with the structure.

3.3.4 Wood-boring beetles

There are a number of beetles (and a few other wasps, bees and weevils) that may infest wood. While most of these attack live trees (and are thus outside of the scope of this chapter), there are some that attack structural timber. For most of these, it is the larvae that cause the damage as they burrow through the wood. In some cases, the damage due to these insects is mostly cosmetic, but in other cases more significant damage may occur. For instance, beetles of the *Lyctus* family infest only hardwoods and are most commonly found in flooring and wood trim. On the other hand, some powder-post beetles of the *Anobiid* type may infest both hardwoods and softwoods and are most commonly found in moist substructures, such as damp crawl spaces. On the whole, however, these insects make relatively little economic impact.

3.4 Marine borers

Various kinds of marine borers will attack wood in saltwater or in brackish water. They are found in all of the seas and oceans, but are most active at temperatures of about 10°C.

3.4.1 Shipworms

Shipworms are wormlike molluscs, related to clams and oysters, the most destructive of which are the *teredos*. They are a particular problem in wharves and harbours, where they attack pilings, docks and so on; in extreme cases, they can destroy structural timbers within a year. They tunnel into the wood, creating tunnels up to 12 mm in diameter and up to 1 m in length. They live on the wood borings and the organic material contained in seawater.

3.4.2 Wood lice

Wood lice (Fig. 12) are crustaceans related to crabs and lobsters. They too bore into the wood, but unlike teredos, they confine themselves to the region just below the wood surface. The damage is greatest in the intertidal zone, particularly as the regions weakened by the small burrows are further damaged by wave action, and by abrasion from floating debris.

4 Non-biological factors that cause deterioration

4.1 Weathering

The weathering of wood is a complex process, involving drying and wetting effects, exposure to light, freezing and thawing, and exposure to chemicals. Weathering is primarily a surface effect, and so does not particularly affect the mechanical

Figure 12: Wood louse.

properties of wood. It is thus mainly of concern with regard to the appearance of the wood.

When freshly cut wood is exposed to the atmosphere, the first noticeable effect is a change in colour, first to yellow or brown, and then to the commonly observed grey due to some chemical breakdown of the cellulose. Beyond this point, in the absence of biological attack, the colour will remain largely unchanged. Exposure to light, particularly ultraviolet light, will increase the severity of the chemical changes. Due to cycles of shrinking and swelling as the wood is dried and rewetted, and sometimes to cyclic freezing and thawing, the surface fibres may loosen and wear away, leading to a very slow surface *erosion* (estimated to be perhaps 6 to 12 mm/century).

In addition, because of non-uniform moisture changes, the surface will roughen, the wood may tend to warp, and cracks may develop. Weathered wood may also be rather more susceptible to biological attack.

4.2 Exposure to chemicals and other environmental factors

In general, wood is quite resistant to *chemical attack*. However, wood may be affected, sometimes seriously, by exposure to certain chemicals. Above the fibre saturation point, changes in moisture content have little effect on wood properties. However, below this point, most mechanical properties of wood are related to the moisture content, as indicated by the following empirical relationship:

$$\log P = \log P_{12} + [(M - 12)/(M_p - 12)] \log (P_g/P_{12}),$$

where $P =$ property of interest, $P_{12} =$ value of property at 12% moisture content, $M =$ moisture content, $M_p =$ moisture content at fibre saturation point (often taken as 25%), $P_g =$ value of property for moisture contents above M_p, i.e. for green timber. Thus water, alcohol and certain other organic liquids that cause the wood to swell, but which do not otherwise alter the wood structure, may bring about completely *reversible* changes in wood properties, in accordance with the above equation. (It should be noted that this empirical relationship does not hold well for impact, bending, toughness or tension perpendicular to the grain). That is, when these liquids are removed, the wood will return to its original state.

Other chemicals, however, may cause *irreversible* changes to the wood structure, and degrade its mechanical properties. Thus, while wood is quite resistant to weak acids, it will be attacked by strong acids, which may attack the cellulose or hemicellulose. Highly acidic salts will also attack wood. On the other hand, alkalis and alkali salts will react with the cellulose and the lignins, thus weakening the wood. In general, hardwoods are more susceptible than softwoods to both acid and alkali attack. Iron salts, from the corrosion of iron fittings such as tie plates, bolts and so on, may degrade damp wood, leading to both the commonly observed discolouration around such fittings, accompanied by a softening of the wood. Of course, since different chemicals will have different effects on different species of wood, it is not possible to make more sweeping generalisations about the resistance of wood to chemicals.

High levels of *nuclear radiation* are known to degrade the strength of wood by decreasing the degree of polymerisation of the cellulose molecules, but this is unlikely to be much of a problem in practice, since nuclear reactor vessels would never be constructed of wood.

Thermal effects on wood properties are more significant. Below approximately 200°C, the mechanical properties of wood increase in an essentially linear fashion as the temperature decreases. If wood is heated or cooled below about 65°C fairly rapidly, the changes in properties with temperature are reversible. However, if wood is exposed to higher temperatures than that for long periods of time, the wood itself will degrade, probably due to acid hydrolysis of the cellulose. There will be a loss not only in strength, but also in mass. This effect is more marked in hardwoods than in softwoods. The loss in properties is also greater at higher moisture contents. Thermal effects can be quite dramatic. For instance, keeping wood at 115°C for 250 days will reduce its strength by about 35%; at 135°C, this change will occur in about 50 days; and at 175°C, it will occur in less than one day.

5 Fire

Wood is obviously a highly flammable material; it has been used as a source of fuel for at least 500,000 years. However, because of the way in which wood burns, timber structures may have a remarkable degree of fire resistance, sometimes even better than that of steel structures. The fire resistance of large sections of wood depends on temperature, moisture content, duration of heating and specimen geometry.

The *ignition temperature* for wood is not a well-defined property; it depends both on the way in which it is measured, and on the type of wood. The value of 200°C is often taken as the temperature at which wood will ignite in the presence of an open flame (which will ignite the combustible gases given off as the wood is heated), but this depends on the duration of exposure to that temperature. For instance, after about 30 minutes of exposure, many woods will ignite at about 180°C; at 300°C they will ignite in about 2 minutes; and at 400°C, in 30 seconds or less. If wood

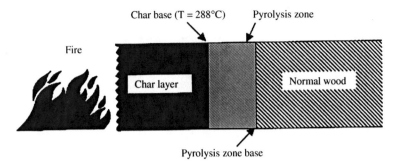

Figure 13: Degradation of wood exposed to fire on one surface.

is simply heated in hot air (in the absence of a flame), ignition may occur at temperatures as low as 330°C in less than one hour.

The particular way in which wood burns is the reason that large wood timbers are remarkably fire-resistant. Fig. 13 shows schematically the degradation zones in a large wood member exposed to fire on one face only. When burning begins, a *char layer* develops on the surface, which slowly progresses into the wood. At the base of this layer, the temperature is about 288°C. However, since wood is an excellent insulator, the temperature falls off rapidly beyond the char base; for Douglas fir, the temperature would be only about 93°C, 13 mm in from the char base. In the *pyrolysis zone*, the wood begins to decompose at a temperature of about 66°C, emitting water vapour, smoke, and both flammable and non-flammable gases. The char zone itself is also a very good insulator, and this helps to prevent strength loss in the interior of the timber section. The rate at which the char base moves into the wood depends upon the species, the density and the moisture content. It is slower at higher moisture contents and densities, and for more impermeable species of wood. For dry Douglas fir, the charring rate is only about 0.6 mm per minute in large sections.

Of course, the spread of a fire through a structure depends on much more than the properties of the wood itself; the entire structural system must be properly designed to provide fire protection. This would include proper fire separations both horizontally and vertically, properly protected openings for stairs and doors, avoidance of combustible finishings and furniture, proper protection for light framing and so on. A detailed discussion of design procedures for fire is well beyond the scope of this chapter.

6 Techniques for preserving wood

Over the years, many different techniques have been developed for wood preservation, depending on the type of wood, the nature of the problem, the exposure conditions and costs. Often, this may consist only of regular painting and maintenance. However, there are many instances in which chemical preservatives must be applied in order to protect the wood from the types of deterioration discussed earlier.

Both the *amount* of preservative that is impregnated into the wood and the *depth of penetration* must be considered when assessing the relative effectiveness of the different types of treatments. These factors will depend, in turn, on the method by which the preservative is applied. If *non-pressure* methods of application are used, such as brushing, spraying or dipping, there will not be much absorption or penetration of the preservative into the wood. Much more effective treatment can be obtained by the application of *pressure* (sometimes in conjunction with the application of a vacuum) to force a larger quantity of the preservative deeper into the wood. The chemical preservatives that are used may be dissolved either in water or oil. In all cases, however, the preservatives must not be unduly flammable or harmful to people or animals, and they must be economical to use.

6.1 Waterborne preservatives

Waterborne preservatives consist of salts that are dissolved in water. They are used largely for timber and plywood in general residential and commercial structures. They are becoming increasingly popular because they are essentially odourless and because they leave a clean wood surface that may be painted or stained. In North America, the most common of these is *chromated copper arsenate* (CCA) that is available in three different combinations of chromium trioxide, copper oxide and arsenic pentoxide, depending on the specific application. It is now often used for poles, posts, pilings and foundation timbers as it provides protection against fungi and termites. Some environmental concerns have, however, been raised about its use. It has been found that in some areas, the CCA may drain off freshly treated wood, or leach out of treated wood into the soil, groundwater or nearby rivers and streams, leading to elevated concentrations of arsenic.

Several other waterborne preservatives are also recommended by ASTM. *Acid copper chromate* (ACC) is a combination of copper sulphate and sodium dichromate with some chromic acid, and makes wood more resistant to termites and decay. *Ammoniacal copper arsenite* (ACA) is often used for timber and plywood intended for house foundations. It consists of copper and arsenic salts in an aqueous ammonia solution. *Chromated zinc chloride* (CZC) can provide protection not only against insects and decay, but also against fire. It consists of a combination of zinc chloride and sodium dichromate. It is best used in dry conditions, as it can be leached out of wood in a moist environment. *Fluor chrome arsenate phenol* (FCAP) consists of fluoride (either sodium or potassium), sodium dichromate, sodium arsenate and dinitrophenol. It, too, performs best above ground as, like CZC, it is subject to leaching.

6.2 Oilborne preservatives

By far the most widely used of the oilborne preservatives is *creosote*, produced mostly from the distillation of coal tar. It is highly toxic to the fungi and insects that attack wood, is insoluble in water and has great permanence because of its low volatility. It is also relatively inexpensive and can easily be applied. It is particularly effective on large timbers, such as railroad ties (sleepers) and bridge timbers. Unfortunately, it has an unpleasant odour and leaves an oily surface that prevents the wood from being painted. It is thus unsuitable for most residential construction. Creosotes can also be manufactured from other organic materials, but such creosotes tend to be less effective than those made from coal tar.

Coal tar creosote may be mixed with coal tar or petroleum to produce *creosote solutions*, mostly in order to reduce the cost of the preservative. They are used in the same way as creosote and are generally almost as effective. Indeed, they are better than creosote in providing protection against checking and weathering of the wood.

Pentachlorophenol (C_6Cl_5OH), often referred to simply as 'Penta', can be dissolved in various mineral spirits and petroleum oils. It is then toxic to wood-destroying organisms. It too leaves a surface that cannot generally be painted.

Copper naphthalene and *tributyltin oxide* have been found to offer reasonable protection against marine borers.

Copper-8-quinolinolate is recommended where the treated wood is to be used for harvesting, storage or transportation of foods.

6.3 Weather-resistant coatings

In order to protect wood against weathering, a large number of surface coatings are available. These include everything from water-repellent treatments to pigmented stains, varnishes and paint. Of these, paints are most effective and are generally used not only for preservation but also for aesthetic reasons. Of course (as any homeowner knows), all of these surface treatments have a relatively short effective life, and so the wood must be refinished on a regular basis for them to retain their preservative qualities.

7 Fire retardants

As noted in Section 5, a properly designed and constructed timber structure can be remarkably fire-resistant. In addition, to satisfy most fire codes, the wood should also be chemically treated with one of the many available fire retardants. While none of the chemical treatments can render wood truly 'fire-proof', they can considerably improve its fire-retarding capacity. The mechanisms by which fire retardants act are still not fully understood. It is believed that some retardants provide a coating that insulates wood as well as prevents air and flame from coming into contact with the wood itself. Some retardants may also provide a gas that inhibits the combustion of the gases given off by the wood as it is heated. Retardants may as well change the thermal reactions that occur during combustion. Depending on the particular situation, and on the particular fire retardant employed, any or all of these mechanisms may occur.

There are two general types of fire-retardant treatments: surface coating with fire resistant paint and pressure impregnation with waterborne salts.

7.1 Surface coatings

While surface coatings are generally less effective than pressure impregnation, they provide the only means of providing some protection to existing structures. The effectiveness of these paints depends on their chemical composition and on the thickness of the applied coating. A number of different formulations are available. In water-based paints, ammonium phosphate or sodium borate are commonly used to provide some fire resistance. In oil-based paints, chlorinated paraffins and alkyds plus antimony trioxide are most common. Most of these fire-retarding paints are intended for interior use, though there are a few that may be used on exterior surfaces.

7.2 Pressure impregnation

Like surface coatings, most pressure impregnation treatments are also intended for interior use, though again there are a few formulations for outdoor exposure. Because of the depth of penetration and the amount of chemicals that may be forced into the wood, this technique provides a greater degree of protection than do surface coatings. To be effective, the amount of fire-retardant salt which penetrates into the wood should be in the range of 40–80 kg/m³. It should be noted, however, that these fire-retardant treatments generally have some effect on the mechanical properties of wood. Strength may be decreased by up to 20%; in the United States, the allowable design stresses for treated wood are reduced by 10% compared with those for untreated wood. There may be some difficulties in machining treated wood because of the abrasive effects of the salt crystals that are used in most fire retardants on the cutting tools. As well, gluing of treated wood becomes more difficult.

The salts that are typically used are combinations of ammonium sulphate, zinc chloride, monoammonium phosphate, diammonium phosphate, boric acid and sodium tetraborate. The salts are best impregnated into air-dry or kiln-dry wood, and a depth of impregnation of at least 1.3 cm should be achieved. It should be noted that since most of the salts used for this purpose are water-soluble, care must be taken to ensure that they are not leached back out of the wood by exposure to excess moisture.

8 Economic considerations

It is difficult to over-estimate the cost to individuals, and to society at large, of permitting wood to rot or otherwise degrade prematurely. Globally, the annual losses due to decay of wood from the biological processes discussed earlier has been estimated to be US$ 10 billion. For instance, in the United States alone, homeowners spend at least US$ 500 million per year simply on replacing rotted and termite damaged wood, exclusive of labour costs. Similarly, in the Canadian province of

British Columbia alone, the cost of repairs to the 'leaky condominiums' built in the late 1980s and early 1990s was estimated to be up to US$ 500 million. The cost of 'doing it right' in terms of both the treatment of the wood itself and the architectural design (rain screens, protective envelopes, proper attention to ventilation, etc.) might have added 7%–10% to the original selling price of each unit, but would have saved vastly more in terms of both reconstruction costs and personal suffering.

Literature for further reading

[1] Dietz, A.G., Schaffer, E.L. & Gromala, D.S. (eds.), *Wood as a Structural Material*, Vol. II, Educational Modules for Materials Science and Engineering (EMMSE) Project, Materials Research Laboratory, The Pennsylvania State University, University Park, PA, 1982.

[2] Morris, P.I., Understanding biodeterioration of wood in structures, Document # W-1507, Forintek Canada Corp., Vancouver, BC, 1998.

[3] Morris, P.I., Wood preservation in Canada – 1999. *Mokuzai Hozan*, **25(4)**, pp. 2–12, 1999.

[4] US Department of Agriculture, *Wood Handbook – Wood as an Engineering Material*, Forest Service, Forest Products Laboratory, Madison, WI, 1999.

[5] Canadian Forestry Service, *Canadian Woods: Their Properties and Uses,* 3rd edn, University of Toronto Press: Toronto, 1981.

[6] *Wood Structures – A Design Guide and Commentary*, Committee on Wood, ASCE Structural Division, American Society of Civil Engineers, New York, NY, 1975.

[7] Wangaard, F.F. (ed.), *Wood: Its Structure and Properties*, Vol. I, Educational Modules for Materials Science and Engineering (EMMSE) Project, Materials Research Laboratory, The Pennsylvania State University, University Park, PA, 1981.

[8] Nichols, D.D. (ed.), *Wood Deterioration and Its Prevention by Preservative Treatments: Degradation and Protection of Wood*, Syracuse University Press: Syracuse, NY, 1982.

Index

The Great Structures in Architecture

F. P. ESCRIG, Universidad de Sevilla, Spain

Starting in antiquity and finishing in the Baroque, this book provides a complete analysis of significant works of architecture from a structural viewpoint. A distinguished architect and academic, the author's highly illustrated exploration will allow readers to better understand the monuments, get closer to them and to explore whether they should be conserved or modified.

Contents: Stones Resting on Empty Space; The Invention of the Dome; The Hanging Dome; The Ribbed Dome; A Planified Revenge - Under the Shadow of Brunelleschi; The Century of the Great Architects; The Omnipresent Sinan; Even Further; Scenographical Architecture of the 18th Century; The Virtual Architecture of the Renaissance and the Baroque.

Series: Advances in Architecture Vol 22
ISBN: 1-84564-039-X 2005 272pp
£95.00/US$170.00/€142.50

Structural Studies, Repairs and Maintenance of Heritage Architecture IX

Edited by: C. A. BREBBIA, Wessex Institute of Technology, UK, A. TORPIANO, University of Malta, Malta

This book contains most of the papers presented at the Ninth International Conference on Structural Studies, Repairs and Maintenance of Heritage Architecture. The Conference was held in Malta, a state smaller than many of the cities that this Conference has visited, and yet that is packed, in the full meaning of the word,

with a history of heritage architecture that spans nearly six millennia - as far as we currently know!

The islands of Malta have limited material resources, in fact, only one - limestone, and a rather soft one at that. However, out of this resource, our ancestor builders have fashioned the habitat for their lives, as these unfolded and changed over the centuries. The problems and efforts that are being made to repair, restore, conserve and protect such limestone architectural heritage are considerable and mirror similar problems faced by other architects, engineers, curators, art historians, surveyors and archaeologists in other countries throughout the world.

The papers featured are from specialists throughout the world and divided into the following topics: Heritage architecture and historical aspects; Structural issues; Seismic behaviour and vibrations; Seismic vulnerability analysis of historic centres in Italy; Material characterisation; Protection and preservation; Maintenance; Surveying and monitoring; Simulation modelling; and Case studies.

WIT Transactions on The Built Environment, Volume 83
ISBN: 1-84564-021-7 2005 672pp
£235.00/US$376.00/€352.50